IPHONE 14
USER GUIDE

AN EASY, STEP-BY-STEP GUIDE ON MASTERING THE USAGE OF YOUR NEW IPHONE 14. LEARN THE BEST TIPS & TRICKS, AND DISCOVER THE MOST USEFUL SECRETS TO GET THE MAX OUT OF YOUR DEVICE

**BY
ETHAN COPSON**

IPHONE 14 USER GUIDE © COPYRIGHT 2022

ALL RIGHTS RESERVED

ETHAN COPSON

Disclaimer

This document is intended to provide accurate and reliable information about the topic and issue at hand. The publication is offered with the understanding that the publisher is not obligated to provide accounting, legally allowed, or otherwise qualifying services. If legal or professional counsel is required, a qualified practitioner should be consulted.

- From a Declaration of Principles that was adopted and approved equally by an American Bar Association Committee and a Publishers and Associations Committee.

It is not legal to reproduce, transmit or duplicate any part of this text in electronic or printed form. The recording of this publication is completely prohibited, and storing of this material is not permitted unless prior permission from the publisher is obtained. Every right is reserved.

The material offered herein is proclaimed to be truthful and consistent in that any liability coming from the use or misuse of any methods, policies, or directions contained within, whether due to inattention or otherwise, is wholly and completely the responsibility of the receiving reader. Under no circumstances will the publisher be held legally responsible or liable for any compensation, damages, or monetary loss caused directly or indirectly by the material contained herein.

All copyrights not held by the publisher are owned by the respective authors.

The material provided below is provided primarily for informational reasons and is therefore universal. The information is presented without any type of contract or guarantee assurance.

The trademarks utilized are used without authorization, and the trademark is published without permission or backing from the trademark owner. All trademarks and brands mentioned in this book are merely for clarification purposes and are owned by the owners, who are unrelated to this document.

TABLE OF CONTENTS

INTRODUCTION 1

CHAPTER ONE: APPLE IPHONE HISTORY 6

CHAPTER TWO: THE IPHONE 14 AND IPHONE 14 PLUS 24

CHAPTER THREE: THE IPHONE 14 PRO AND IPHONE 14 PRO MAX 35

    7 BIG CHANGES 47

CHAPTER FOUR: THE EASIEST WAY TO SET UP YOUR NEW IPHONE 14 ... 52

CHAPTER FIVE: HOW TO USE SIRI ON IPHONE 14 56

CHAPTER SIX: HOW TO CLOSE OR CLEAR APPS ON IPHONE 14 64

CHAPTER SEVEN: HOW TO FORCE RESTART IPHONE 14 67

CHAPTER EIGHT: HOW TO ENABLE AND ACTIVATE IMESSAGE ON IPHONE 14 69

CHAPTER NINE: TAKING A SCREENSHOT ON AN IPHONE 14 72

    TAKE A SCREENSHOT USING PHYSICAL BUTTONS 72

    USE ASSISTIVE TOUCH TO TAKE SCREENSHOT 74

CHAPTER TEN: APPLE'S DYNAMIC ISLAND 77

    WHAT IS APPLE'S DYNAMIC ISLAND? 77

    WHAT DOES DYNAMIC ISLAND DO? 78

    WHICH APPS SUPPORT DYNAMIC ISLAND? 78

    HOW DOES DYNAMIC ISLAND WORK? 79

    WHICH PHONES ARE COMPATIBLE WITH DYNAMIC ISLAND? 82

CHAPTER ELEVEN: THE IPHONE 14 CAN CONNECT TO SATELLITES 83

    HOW TO SOLVE THE SATELLITE ISSUE 83

    FOR EMERGENCY SERVICES AND FRIENDS 84

    HOW MUCH WILL SOS VIA SATELLITE COST? 85

CHAPTER TWELVE: HOW CRASH DETECTION WORKS AND HOW TO TURN IT OFF 87

    CRASH DETECTION: WHAT IS IT? 87

    HOW DOES CRASH DETECTION WORK? 87

TABLE OF CONTENTS

WHAT KINDS OF CRASHES CAN IT DETECT? 88

HOW TO ENABLE CRASH DETECTION 89

CHAPTER THIRTEEN: HOW TO USE APPLE PAY 90

SETTING UP APPLE PAY 90

USING APPLE PAY 92

PAY USING AN IPHONE OR APPLE WATCH 93

SEND MONEY WITH MESSAGES 93

CHAPTER FOURTEEN: HOW TO MASTER THE IPHONE 14 PRO & IPHONE 14 PRO MAX CAMERA 97

VOLUME BUTTONS 97

CONTROLLING THE ZOOM 98

APPLE QUICKTAKE 99

BONUS CONTROLS 99

VIDEO RECORDING ON IPHONE 14 PRO 100

CHAPTER FIFTEEN: IOS 16 102

IOS 16: FEATURES 103

1) Add widgets to multiple lock screens *103*

2) Change how lock screen notifications are shown *103*

3) Edit and unsend messages *104*

4) Restore just-deleted texts *104*

5) Set up filters in Focus mode *105*

6) Schedule the sending of emails *105*

7) Lock down your iPhone *105*

8) Share tab groups in Safari *105*

9) See the battery level percentage *106*

10) Plan routes with multiple stops in Apple Maps *106*

11) Track your fitness without an Apple Watch *106*

12) Run a Safety Check *107*

13) Cut out objects from photos *107*

14) Track your medications *107*

.... *108*

15) Add haptic feedback to the keyboard *108*

16) Share photos more easily *108*

HOW TO GET IOS 16 ON YOUR OLD IPHONE 108

How to customise your iPhone lock screen wallpaper, font, colour, and widgets 109

    HOW TO CREATE A CUSTOM IPHONE LOCK SCREEN 110
    EDIT A LOCK SCREEN 110
    DELETE A LOCK SCREEN 111
    SWAP OUT THE LOCKS ON THE DOORS 111
    HOW TO ADD A PHOTO TO YOUR IPHONE LOCK SCREEN 111
    HOW TO ADD WIDGETS TO YOUR IPHONE LOCK SCREEN 112
    HOW TO ADD A FOCUS TO YOUR IPHONE LOCK SCREEN 112
    HOW TO CUSTOMIZE YOUR IPHONE HOME SCREEN AESTHETIC 113
    OTHER IOS 16 TRICKS 113
    HIDDEN IOS 16 FEATURES 114

CHAPTER SIXTEEN: HEALTH APP WITH IOS 16 118

    MONITORING MEDICATIONS 118
    MEDICATION LOGGING 121
    Apple Watch Medication App 122
    MEDICATION INTERACTIONS 122
    MEDICATION SIDE EFFECTS AND OTHER INFO 123
    EXPORT MEDICATIONS 124
    SLEEP FEATURES 124
    HEALTH SHARING INVITATIONS 125

CHAPTER SEVENTEEN: SETTING UP FITNESS APP WITH IOS16 126

    ISSUES & TROUBLESHOOTING 131
    UPDATE SOFTWARE 132
    CHECK MOTION CALIBRATION & DISTANCE SETTINGS 132

CHAPTER EIGHTEEN: TRICKS TO BOOST BATTERY LIFE 134

    CHARGING YOUR IPHONE 14 137

CHAPTER NINETEEN: WAYS TO SELL OR TRADE IN YOUR OLD IPHONE 144

CHAPTER TWENTY: YOUR IPHONE 14 PROTECTION 147

    ARE IPHONE 13 CASES COMPATIBLE WITH THE IPHONE 14? 149

CONCLUSION 152

    The Importance of iPhone Maintenance 154

INTRODUCTION

At the Far-Out event, Apple unveiled its new iPhone 14 models. The four versions are as anticipated: the iPhone 14, iPhone 14 Plus, iPhone 14 Pro, and iPhone 14 Pro Max.

We cover all the information you need to know about the new phones in this book, from their release date and price to changes in their designs, technological specifications, and new features.

RELEASE DATE

The iPhone 14 phones were unveiled on September 7 during Apple's Far Out event.

On September 9, all four versions will be available for pre-order. Three of them (the 14, 14 Pro, and 14 Pro Max) will subsequently go on sale on September 16 after a week.

PRICE

The 14 and 14 Plus have $799 and $899 starting prices, respectively, while the 14 Pro and 14 Pro Max have $999 and $1,099 starting prices. The complete pricing list for the common models is provided here:

- iPhone 14 (128GB): $799 / £849 / $1,399
- iPhone 14 (256GB): $899 / £959 / $1,579
- iPhone 14 (512GB): $1,099 / £1,179 / $1,899
- iPhone 14 Plus (128GB): $899 / £949 / $1,579
- iPhone 14 Plus (256GB): $999 / £1,059 / $1,749
- iPhone 14 Plus (512GB): $1,199 / £1,279 / $2,099

Here are the costs of the pro versions:

- iPhone 14 Pro (128GB): $999 / £1,099 / $1,749
- iPhone 14 Pro (256GB): $1,099 / £1,209 / $1,899
- iPhone 14 Pro (512GB): $1,299 / £1,429 / $2,249
- iPhone 14 Pro (1TB): $1,499 / £1,649 / $2,599
- iPhone 14 Pro Max (128GB): $1,099 / £1,199 / $1,899
- iPhone 14 Pro Max (256GB): $1,199 / £1,309 / $2,099
- iPhone 14 Pro Max (512GB): $1,399 / £1,529 / $2,419
- iPhone 14 Pro Max (1TB): $1,599 / £1,749 / $2,769

The current iPhones will be made available for pre-order on Apple's website.

DESIGN ADJUSTMENTS

Here we have two quite distinct tales. Although you do now have the choice of a bigger screen, which we'll examine in due time, the iPhone 14 is virtually similar to the iPhone 13; however, the iPhone 14 Pro and Pro Max have undergone major design modifications.

The two smaller floating apertures for the front-facing sensors replace the notch on the Pro versions, as planned, but they have been given far more prominence than was anticipated.

Apple refers to this as the Dynamic Island since it is connected by a black bar and resembles a single, narrow lozenge. Depending on the program and environment, its appearance, including size and controls, changes. It functions as a little widget that shows running times, sports scores, and other essential data from apps you've swiped away or allows you to interact with apps without fully opening them.

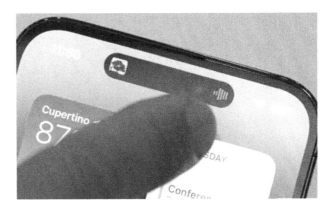

We have to wait at least another year for an iPhone design that is really all-screen, but this cleverly turns that inconvenience into a positive.

The camera module on the back of this year's Pro models is bigger than it was last year, but the bezels surrounding the screen are thinner this year. The actual lenses are larger in diameter and protrude more from the phone's rear. Although doing so would prevent them from enjoying the attractive new color finishes, we anticipate that the majority of purchasers will wish to cover this protruding module by placing the 14 Pro in a case.

COLOR OPTIONS

Five colors are available for the 14 and 14 Plus: Midnight, Starlight, (Product) Red, blue, and a brand-new finish in purple. The Pro variants are available in silver, gold, the new Space Black, and Deep Purple, a distinct and considerably deeper purple.

SCREEN UPGRADES

The 5.4-inch iPhone mini format does not have a new iteration. (However, shoppers with little hands still have access to the iPhone 13 mini on sale.) Apple has instead introduced the iPhone 14 Plus, a large-screen non-Pro variant.

Thus, this is the updated lineup:

- iPhone 14 (with 6.1-inch screen)
- iPhone 14 Plus (with 6.7-inch screen)
- iPhone 14 Pro (with 6.1-inch screen)
- iPhone 14 Pro Max (with 6.7-inch screen)

In comparison to the new phone offerings from the previous fall:

- iPhone 13 mini (with 5.4-inch screen)
- iPhone 13 (with 6.1-inch screen)
- iPhone 13 Pro (with 6.1-inch screen)
- iPhone 13 Pro Max (with 6.7-inch screen)

ALWAYS-ON DISPLAY

The displays of the non-Pro variants are virtually the same as those of the 13 except for the size of the 14 Plus. However, the 14 Pro and 14 Pro Max have a screen that is constantly on. When the device is inactive, this will automatically dim and reduce animation and onscreen power consumption (it may dynamically decrease to a refresh rate of only 1Hz), while still displaying the most crucial data and certain widgets.

Because of the always-on display, you can view important information even when the screen is dull and yet save battery life.

With 1,600 nits of peak HDR brightness and up to 2,000 nits outside, the 14 Pro display is also brighter than the 13 Pro's.

CAMERA UPGRADES

This year's iPhones both get camera enhancements, but it's impossible to tell which is more amazing.

NON-PRO MODELS

The 14 and 14 Plus iPhones will be discussed first.

Similar to the previous year, this device has twin cameras with 12MP wide and ultra-wide lenses as its core specifications. However, Apple claims that the primary camera's quicker aperture and bigger sensor will result in greater performance.

Low-light photography is a significant aspect of the 14's photographic advancements; according to Apple, it performs 49% better under these circumstances. Additionally, a new Action Mode has been added that is intended to enhance video stabilization.

Additionally improved is the front-facing TrueDepth camera. It gains autofocus, and you should also notice better low-light performance in this area because of a larger aperture.

PRO MODELS

Here, the main enhancement was far in advance disclosed. The iPhone 14 Pro now has a 48MP wide-angle camera on the back, as was to be anticipated.

However, that outstanding number can be a little deceptive. The majority of the pictures you take with the 14 Pro won't even be 48MP; they'll be 12MP, but they'll have far more information since the camera employs image processing to turn four pixels into one.

This sensor is 65 percent bigger than the Wide camera in the iPhone 13 Pro, which guarantees greater low-light performance, again up to two times better, according to Apple, in addition to having more megapixels. A new Photonic Engine computational technology is also available to assist in capturing greater detail in low-light conditions.

The camera module in the iPhone 14 Pro is larger than ever, but this is justified by an improved feature and spec list.

SPECIFICATIONS AND NEW FEATURES

Under the hood, there have been several advancements. Here are the main updates to the specs and new features.

SAFETY FEATURES

A new accident detection capability and satellite connection for emergencies are the two key facets of Apple's increased emphasis on personal safety, both of which are included in the new Apple Watches.

Crash detection is similar to the earlier Fall Detection function in that it utilizes an updated gyroscope and accelerometer to identify whether you are inside a car that is colliding, but it is potentially far more hazardous. The iPhone will contact emergency services and alert your selected contacts if it believes you have been in an accident and will do so until overridden.

In contrast, Emergency SOS via satellite is a function that enables you to contact emergency services while you are outside of cellular coverage via satellite communication. Although it relies on brief pre-written messages, a team of intermediates will take care of it for you if the emergency provider you're attempting to reach only takes calls. The function will show you where to aim the device in order to create and maintain a strong connection, but you will need a clear view of the sky.

Only the US and Canada will initially have access to the satellite function. Although Apple claims it would be free for the first two years after purchase, it is probable that it will eventually have an ongoing charge.

NEW A15 AND A16 PROCESSORS

Although the term "new" is dubious, there has been some development in the processor area.

All four of the 13-series iPhones from last autumn still use A15 Bionic CPUs, exactly like the non-Pro variants. However, this has been modified to include a second GPU core (similar to the one in the 13 Pro from last year), and should thus run considerably better than the basic 13. (It's also conceivable that, as had been suggested before to the presentation, Apple has upped the RAM allotment from 2021. Apple doesn't make RAM information available to the public, so we won't know until review samples start to arrive.)

Much more intriguingly, the Pro versions get A16 Bionic CPUs from the next generation. This is being produced using a 4nm rather than a 5nm technology, as was projected back in 2020, which should lead to increased performance and power efficiency. While Apple hasn't provided many specifics about the performance we can anticipate from the A16 (other than the fact that it has 16 billion transistors and the Neural Engine can process nearly 17 trillion operations per second, up from 15.8 trillion in the A15), the company has stated that the A16 is 40% faster than the closest rival.

That is to say, until the existing software catches up with its capabilities, the 14 Pro will completely destroy all speed benchmarks while providing less visible increases in real-world performance.

BATTERY LIFE

Apple claims that the battery life has improved somewhat across the board, with the three non-Plus versions all gaining an hour over the previous iteration. Of course, the iPhone 14 Plus lacks a direct predecessor, but because of its bigger body and larger battery, it should easily surpass the iPhone 14 Pro. These are the estimated numbers:

- iPhone 14 Pro Max: Has 29 hours of video playback (up from 28 hours on 13 Pro Max)
- iPhone 14 Pro: Has 23 hours (up from 22 hours on 13 Pro)
- iPhone 14 Plus: Has 26 hours
- iPhone 14: Has 20 hours (up from 19 hours on 13)

CHAPTER ONE
APPLE IPHONE HISTORY

It's difficult to believe now, but the first iPhone did not have many of the functions we take for granted today, like copy and paste, 3G—certainly not 5G—or even applications. Heck, you were limited to using iTunes on a PC to sync it.

The tale of the Apple iPhone begins in 2007, when Steve Jobs shocked the world by revealing a mystery 3-in-1 gadget called the first Apple iPhone, which for the first time combined an iPod music player, an Internet communication device, and a regular phone.

And although while the iPhone has undergone significant development since its creation, the initial structure was sound and guaranteed success for years to come. Every iPhone would have consistent, fluid performance, a recognized, user-friendly interface, an emphasis on privacy, and a unique collection of features.

IPHONE 14 SERIES: PRICE, RELEASE DATE, FEATURES, AND SPECS

In this history, we take a look at the innovation that came with each new iPhone, and all the major features that each model brought, starting from the very first one and going to the most recent phones. Join us for this ride right below.

Apple iPhone History:

- iPhone (year 2007)
- iPhone 3G (year 2008)
- iPhone 3GS (year 2009)
- iPhone 4 (year 2010)
- iPhone 4s (year 2011)
- iPhone 5 (year 2012)
- iPhone 5s and iPhone 5c (year 2013)
- iPhone 6 and iPhone 6 Plus (year 2014)
- iPhone 6s and iPhone 6s Plus (year 2015)
- iPhone SE (year 2016)
- iPhone 7 and iPhone 7 Plus (year 2016)
- iPhone 8 and iPhone 8 Plus iPhone X (year 2017)
- iPhone XR, XS and XS Max (year 2018)
- iPhone Pro Max, iPhone 11 Pro and iphone11 (year 2019)
- iPhone SE (year 2020)
- iPhone 12 mini, iPhone 12 Pro, and 12 Pro Max (year 2020)

- iPhone 13 mini, 13, iPhone 13 Pro and iPhone 13 Pro Max (year 2021)
- iPhone SE (year 2022)
- iPhone 14 pro max, 14 Plus, 14 Pro and iPhone 14 (year 2022)

ORIGINAL IPHONE, JUNE 2007

a 412MHz ARM CPU and 3.5-inch screen with 480 x 320 resolution for 163ppi. For the iPhone, this was the beginning of everything. Although it wasn't the first in many aspects, the debut of the smartphone was perhaps the most significant.

"iPhone integrates three items into one compact and lightweight handheld device: a groundbreaking mobile phone, a widescreen iPod with touch controls, and a breakthrough Internet communications device with desktop-class email, online browsing, maps, and searching."

The first iPhone was the catalyst for the development of contemporary smartphones. While there had been smartphones before it, Apple's iPhone surpassed them with its noticeably larger screen, ground-breaking multi-touch interface, and the first functional on-screen keyboard. Although Apple considered the first iPhone to be a phone first, iPod second, and a "communicator" third, it's interesting to note that most people would likely rate the "communicator" feature as what distinguishes the iPhone and smartphones from other devices.

Here is a very brief explanation because a book (or even an article) wouldn't be able to adequately cover the enormous number of ground-breaking breakthroughs in the original iPhone and the history that surrounds each one:

- On-screen keyboard
- 3.5-inch display with a resolution of 320 x 480 pixels
- iOS, based on the multi touch interface controlled entirely by your finger
- 2-megapixel photo camera with a NO video recording capabilities
- Internet Connectivity (2G) with web browser and native email client
- YouTube and Google Maps applications, Google Search
- Ambient light sensor, proximity sensor
- 4GB / 8 GB / 16GB storage models
- iPod music / video player with support for iTunes

IPHONE 3G JULY 2008

Essentially the same as the first iPhone, but with a thinner metallic edge and 3G connection. Take notice of how the App Store icon appears as well. We now utilize our phones differently thanks to the move to centralized app stores.

The App Store went live only one day before Apple's second-generation phone, the iPhone 3G, was introduced on July 10, 2008. Of course, it was made accessible on the first iPhone, and looking back, it is obvious that its release was just as significant as the introduction of the phone itself.

The iPhone 3G maintained the same screen size as the original while adopting a new, glossy plastic body and including 3G connection, which significantly sped up the loading time of online sites. The iPhone 3G was the first iPhone to have GPS, a satellite communication system that enables your phone to know its precise location and is a crucial component of enhanced mapping and navigation.

The 4GB storage option was removed with the iPhone 3G, which only came in 8GB and 16GB variants. The 3.5mm headphone jack on the first iPhone was seriously faulty; it was far too recessed within the phone's body, making it difficult to use many headphones without an adaptor (oh, the irony!). This new iPhone corrects that blunder. In order to guarantee a correct fit, "certain stereo headphones may need an adaptor (available separately)," Apple explicitly acknowledged on the website for the first iPhone.

IPHONE 3GS JUNE 2009

Although it has a faster 600MHz ARM A8 CPU and a fingerprint-resistant screen coating, the design is similar to the 3G. It was about speeding up the experience by adding features like a digital compass and video recording.

The iPhone 3GS delivered significant performance increases and quicker 3G connection, but it was more of an incremental update than a dramatic shift.

The S in 3GS stood for speed since the phone was faster and more potent, but its biggest innovation was perhaps the fact that it was the first iPhone to support video recording. The new 3-megapixel camera on the iPhone 3GS included a video mode and could capture movies at VGA (480p) quality.

Additionally, it was the first iPhone to come equipped with a digital compass, enabling Maps to accurately depict your direction in space.

The remaining innovation was mostly centered on software; fundamental functions like copy and paste, push notifications, landscape keyboard, and others were introduced in 2009.

IPHONE 4 JUNE 2010

With the debut of the Retina display and a 3.5-inch screen with a resolution of 960 x 480, here is where design and power truly took off. With FaceTime, the flattened glass design made way for an iconic front camera.

The iPhone 4 was widely regarded as having the greatest iPhone design ever by Apple users. The iPhone 4 included a striking for the time glass structure with a metal frame as the first significant update inside the iPhone series. It was also the first iPhone to include a "Retina" display, a new 640 x 960 pixel screen resolution that Apple said was so crisp that it was comparable to the retina's natural limitations. For the next several years, this was the only phone with such a high resolution.

Around the iPhone 4, there was also a significant issue known as "antenna-gate." While Apple never officially corrected the problem, it did provide complimentary bumpers that helped. The problem was that when held in a certain position, cellular signal could easily be obstructed by your hand.

Because there are so many improvements on the iPhone 4, we'll simply briefly review the most significant ones below:

- Smaller, micro-SIM card slot
- • The highest phone resolution available at the time, a 'Retina' display
- New glass-and-metal design
- First iPhone with a front-facing, selfie camera
- Secondary mic for noise cancellation
- A 5-megapixel camera capable of 720p HD video recording

The iPhone 4 ultimately debuted on Verizon Wireless at the beginning of 2011 a few months after its introduction on AT&T, Apple's exclusive partner.

IPHONE 4S SEPTEMBER 2011

Similar to the iPhone 4, but with increased speed and the integration of Siri as the personal assistant. CEO Tim Cook unveiled the iPhone 4S on October 4, 2011, the day before Steve Jobs died away on October 5.

The death of Steve Jobs, the man who founded Apple, helped to shape it as a business, and who led it solely with his singular vision for technology, on October 5, 2011, just one day after the release of the iPhone 4s, made 2011 likely the hardest and most difficult year for Apple, its supporters, and the community. Tim Cook, formerly in charge of logistics at Apple, succeeded Jobs as CEO, and Cook presided over the iPhone 4s launch.

The standout feature of the iPhone 4s was Siri, the intelligent voice assistant that dazzled with its aptitude at humorously posing challenging questions and offering consoling humor when you needed it. Alarms and calendar appointments were also scheduled. However, the more recent iPhone still had a 3.5-inch screen, making it smaller than the Android rivals that were gaining popularity with those who preferred larger handsets.

The terrible "antenna-gate" issues of the iPhone 4 were resolved with the release of the two-antenna iPhone 4s. The Apple A5, a more powerful dual-core processor, the new 8-megapixel camera that takes photographs with better white balance and more clarity, and iCloud, a cloud storage option, were the main new features added to the iPhone 4s.

Sprint also obtained the rights to sell Apple's iPhone in October 2011, at which point it began to provide its customers with the iPhone 4s, iPhone 4, and iPhone 3GS.

IPHONE 5 SEPTEMBER 2012

With a bigger 4-inch screen with a 1136 x 640 resolution, Apple made another leap, changing the iPhone's aspect ratio. Additionally, Lightning, a new connection, was introduced.

The Apple iPhone 5 introduced a slightly larger and taller display than previous iPhones, but it was not the drastic change that many users anticipated. The screen size was extended from 3.5 to 4 inches, and a 16:9 aspect ratio (rather than the previous 3:2) was included, which proved more beneficial for video, which is normally filmed in a 16:9 format.

The iPhone 5 was also a watershed event for Apple since it was the first phone to use a processor that Apple had developed independently of Qualcomm. The Apple A5 laid a significant groundwork with its improved design and performance. Now, Apple could better tailor the performance of its phone to the processor it had developed. Additionally, the new iPhone 5 design was lighter and slimmer.

It was the first iPhone to enable 4G LTE, at long last. The color balance for the sRGB standard, which is used across the web for photos and video, was also enhanced.

At the same time, Scott Forstall, the head of iOS development, was also fired because the recently released Apple Maps app, a project he oversaw, started out horribly. Later on, this would cause a significant departure from the skeuomorphic design of the iOS UI.

The last major U.S. carrier finally obtained the rights to sell the iPhone 5 months after its release: T-Mobile joined AT&T, Verizon Wireless, and Sprint in offering the device.

Around this time, John Legere began T-significant Mobile's overhaul and built it to where it is now.

IPHONE 5S AND IPHONE 5C, SEPTEMBER 2013

similar to the iPhone 5 in most respects, but with a plastic body. With a variety of covers to create contrasting patterns, the iPhone 5C was all about color and pleasure.

Everybody waited with bated breath for the Apple in late 2013 since rivals had already made larger-screen phones the norm. Although everyone anticipated a larger phone, the corporation was not yet prepared to reveal it. It has the same size and appearance as the iPhone 5s up its sleeve.

The new Apple A7 "Cyclone" chip, the first 64-bit chip in a phone, years before others had even begun to work on 64-bit chips, and "Touch ID," the fingerprint-based secure identification system that would take years to properly implement on Android phones, were two key features of this important "S" update that were years ahead of the competition. Along with other small upgrades, it also improved the performance of low-light cameras.

The iPhone 5s ushers in a new era for iOS as well. At WWDC in June 2013, Apple revealed iOS 7, which features a fundamental redesign of the iOS user interface. The new interface, created by Jony Ive, did away with the dated features of iOS in favor of a flatter, more transparent interface that seemed like a significant advancement.

The iPhone 5S had the same exterior as the iPhone 5, but it did away with the home button and included Touch ID, which made it possible to unlock the device and verify App Store transactions.

Also, worth mentioning is the iPhone 5c, a colorful, budget-friendly polycarbonate phone that many people purchased.

IPHONE 6 AND IPHONE 6 PLUS, SEPTEMBER 2014

After several years of anticipation, Apple eventually unveiled an iPhone with a larger screen in 2015. The 4.7-inch iPhone 6 and the 5.5-inch iPhone 6 Plus were the two that were available.

It introduced a Galaxy S and Galaxy Note competition, but it was also unmistakably better designed, with a sleek, robust metal body that sold like hotcakes. Compared to Samsung's designs made of cheap-feeling plastic, it felt far better. It's easy to argue that the iPhone 6 was a crucial step towards making premium, metal designs the norm for flagship devices.

A much-improved camera with quicker autofocus and capability for continuous autofocus in films was also included with the iPhone 6 and 6 Plus.

The iPhone 6's basic model was still 16GB, but there were 64GB and 128GB choices for consumers that had higher storage requirements.

IPHONE 6S AND IPHONE 6S PLUS, SEPTEMBER 2015

After being on a campaign for thicker iPhones, Apple actually raised the thickness of its new iPhones for the first time in 2015. The revolutionary display technology that enabled the new iPhones to perceive the form of touch was the primary cause of the thicker body of the devices. Apple gave the technology the name 3D Touch and included it in several of its own applications. Everything functioned somewhat like a right click, a convenient shortcut for many programs to save time.

The first iPhones with 4K video recording capabilities were the iPhone 6s and 6s Plus. Android phones have supported 4K video for years, but they have always had bothersome restrictions like a 5-minute clip time cap and no native 4K editing tools. Although the iPhone 6s was released later than expected, it did everything right: the video was captured at high bit rates, came out incredibly clear, and Apple upgraded the superb iMovie to allow up to two

4K video channels. Many PCs at the time couldn't edit 4K footage, yet a phone could.

Other noteworthy iPhone 6s innovations were supported by the hands-free voice command "Hey, Siri," which activated the intelligent iPhone assistant, a new rose gold color, a set of new 3D Touch-enabled Motion Wallpapers, and Live Photos, which captured a little video before and after a still picture.

IPHONE SE APRIL 2016

The iPhone SE took the rare decision to revert to the iPhone 5 series' earlier design. It accomplished this to provide a more affordable, smaller alternative while retaining the 6S's potent internals, including the A9 CPU and the most recent camera, on a 4-inch display with 1136 x 640 pixels.

To the delight of fans of small phones, the first-generation iPhone SE brought back the recognizable 4-inch form factor from the recent past.

The pricing was what was most astonishing, however; the entry-level 16GB model cost $400, making it the first iPhone that was really cheap.

The SE featured a superb primary camera that performed very similarly to its top 6s contemporaries, and Apple had given it the same potent processor. The battery life, which was inferior to that of the larger versions, was the largest concession that had to be made.

IPHONE 7 AND IPHONE 7 PLUS, SEPTEMBER 2016

Apple did not significantly change the iPhone 7 despite the fact that the debut of Jet Black and the development of (Product) Red generated controversy. The A10 processor powers the iPhone 7, which also adds waterproofing while maintaining the same screen size and quality.

Apple eliminated the headphone port in 2017 and added water resistance to both of the new iPhone models. The 3.5mm headphone jack was destroyed by Apple in a 'courageous,' typical Apple action. Although the iPhone 7 smashed all prior Apple sales records, there was a significant protest, and many people were miffed by this decision, even if it had little impact on sales.

Additionally, the iPhone 7 Plus distinguished itself as the better camera phone for the first time thanks to a dual camera system with a secondary, "telephoto" lens that allowed taking zoomed-in photos and a new "Portrait mode" effect that blurred the background in an image for a polished, DSLR-like appearance. A superior camera was also added to the little iPhone 7; optical image stabilization (OIS) allowed for more blur-free photos and improved video stabilization.

The new jet-black hue, which was only offered in very small amounts and resembled glass but was incredibly easy to scratch, was the second major development. There was also a new matte black shade that was less visually appealing but more useful and less prone to scratches.

By offering a new, wide-color DCI-P3 setting that made everything seem more colorful and stunning, Apple proceeded to enhance the screens of its iPhones.

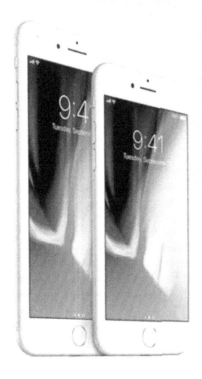

APPLE IPHONE 8 (2017)

The iPhone 8's display size and resolution are likely same as those of the iPhone 7, but Apple introduced True Tone technology, switched back to glass over metal, and increased storage capabilities by a factor of two. Additionally, it updated the CPU to the A11 chip and included support for wireless charging.

APPLE IPHONE 8 PLUS (2017)

The iPhone 8 Plus boasts True Tone technology, wireless charging, a new CPU than its predecessor, and twice the storage of the smaller model. It was the final iPhone model to feature Touch ID as a biometric option, along with the smaller iPhone 8, with the iPhone X setting the bar for future iPhones with Face ID.

APPLE IPHONE X (2017)

In 2017, Apple unveiled a record-breaking three new iPhone models while continuing to sell the iPhone 7, 6s, and SE series, resulting in the company's broadest selection of iPhones ever.

The 10-year anniversary iPhone, the iPhone X, was the largest design change since the 2007 launch of the first model. Additionally, this year was the first time an iPhone was sold for $1,000, a significant increase from the $750 that Apple had previously charged for a Plus-sized iPhone.

And the iPhone X was really the model that laid the way for the future since it was the first to include an OLED display and a bezel-free, edge-to-edge screen (yes, with the notch). The brand-new Face ID system, which identified your face and replaced Touch ID's fingerprint identification technology, proved to be more dependable and user-friendly.

The Apple A11 Bionic processor with its neural engine, the glass back, wireless charging, and the brand-new 64GB and 256GB storage choices were among the other innovations that year that were shared by the iPhone X, iPhone 8, and iPhone 8 Plus.

APPLE IPHONE XR (2018)

Apple continued its practice of releasing three new phones at once in late 2018, but this time, each of the three new devices had Face ID and gesture navigation, which were seen as the company's future direction.

With a price beginning at $750, the iPhone XR was the most reasonably priced model in the line, yet it was still expensive by the standards of typical smartphone pricing. The rear of the gadget received a beautiful new design with brilliant tones with the introduction of the XR. The iPhone XR was the best-selling iPhone of the series thanks to its larger front bezels, LCD screen technology with a relatively low HD+ resolution, and use of the same main camera as on the XS series. These features, along with the appeal of the iOS ecosystem, helped the iPhone XR achieve this success.

APPLE IPHONE XS (2018)

The iPhone XS provided a better camera than the iPhone X, dual-SIM compatibility, the A12 Bionic CPU, 512GB of storage, a longer battery life, and a new Gold finish. However, the 10-year anniversary model's design is still in use.

APPLE IPHONE XS MAX (2018)

A year after the release of the iPhone XS, which celebrated the device's 10th anniversary, Apple unveiled the iPhone XS Max, a larger version of the iPhone X with all the same upgrades.

APPLE IPHONE 11 (2019)

Three new iPhones that Apple released in 2019 were pretty comparable to those it had previously given.

The most widely used model in the line was once again the iPhone 11. The wise decision by Apple to reduce the beginning price of the iPhone 11 to $700—$50 less than the starting price of the iPhone XR from the previous year—helped the device gain more traction. While the competition was ahead in this area, critics noted that Apple was still utilizing an outdated LCD technology and a poor resolution display in a phone that costs quite a bit.

The iPhone 11 replaced the iPhone XR, and although having the same dimensions, it also included a redesigned camera housing with a frosted finish and a second camera on the back. Additionally, there are some gorgeous new hues. The wide-angle camera on the iPhone 11 gives it some fantastic new photographic capabilities over the XR. It also boasts upgraded hardware.

APPLE IPHONE 11 PRO (2019)

With an entirely new, rather divisive camera housing, a frosted matte glass finish, and a ton of technological advancements, particularly in the camera area, the iPhone 11 Pro replaced the iPhone XS.

Apple finally gave consumers what they had been asking for for years in 2019: slightly thicker phones with much larger batteries that propelled the Pro and in particular the Pro Max to the top of the rankings for battery life. The phones also received Night Option, a new camera mode that would intelligently combine a long exposure shot with many others in an automated, simple procedure that would take excellent photos in very low light.

APPLE IPHONE 11 PRO MAX (2019)

The successor to the iPhone XS Max, the iPhone 11 Pro Max, had a similar appearance to the iPhone 11 Pro but on a bigger size. Its 6.5-inch display was stunning, particularly in midnight green, and it had a camera with great low-light performance thanks to Night Mode.

IPHONE SE (2020) APRIL 2020

Early in 2020, when the coronavirus epidemic spread and dealt serious damage to the economy, Apple brought back the SE moniker, and the phone was an immediate smash. It cost $400 and used the most recent Apple A13 processor available at the time, making it far more powerful than any Android phone in that price range.

However, the new iPhone SE (2020) wasn't nearly as little as the 4-inch SE's predecessor. Instead, it really makes use of the chassis from an iPhone 8, giving you a 4.7-inch screen. While other devices in this price range also featured an ultra-wide and sometimes even telephoto lens, the new SE did stand out with the quality of its single lens. The single camera on the rear was capable of shooting 4K films with superb resolution.

The battery life of this phone continues to get the most criticism. Because the internal battery was so little (only 1800mAh), if you used your phone more throughout the day, you would need to recharge it even before you got home from work. This was certainly not a pleasant experience.

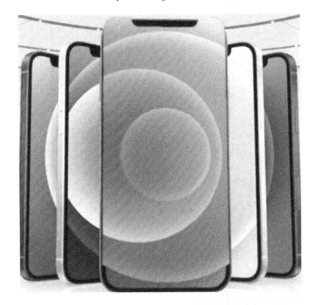

APPLE IPHONE 12 (2020)

In September 2020, Apple unveiled the iPhone 12. It included an updated A14 Bionic processor, a ceramic frame, a 6.1-inch Super Retina XDR OLED display, and better cameras. The iPhone 12 put a lot of emphasis on 5G, with each model having 5G capabilities but only utilizing them when they are available to save energy.

APPLE IPHONE 12 MINI (2020)

A smaller 5.4-inch iPhone model with the same technological capabilities as the bigger one was introduced with the launching of the iPhone 12. Offering the same power to clients who want something portable was the aim.

APPLE IPHONE 12 PRO (2020)

The 6.1-inch iPhone 12 Pro and 6.7-inch Pro Max were part of the 12-lineup. Each of these phones sported an IP68 certification, a Super Retina XDR display, a polished surgical stainless-steel chassis, and, of course, 5G capability.

APPLE IPHONE 12 PRO MAX (2020)

The Apple iPhone 12 Pro Max was the highest model in the iPhone 12 series, with the same screen technology, 5G connection, and CPU as the 12 Pro (although in a different size, of course). The cameras and what they were capable of between the 12 Pro and the 12 Pro Max were the largest difference, however.

As you might expect from Apple, all four 12 series iPhones feature 5G connectivity, a first for any iPhone. This was handled flawlessly with support for mmWave in the US and a clever feature that would automatically switch back to LTE networks when 5G ones were not available to preserve battery life.

Additionally, all four now have OLED screens with gorgeous, deep blacks and rich colors, while the earlier, more cheap iPhone 11 model did it by using an LCD screen.

The best-selling model in the series right now is the iPhone 12, but the Pro iPhones, which are distinguished by their telephoto cameras and cutting-edge photography capabilities like RAW recording, also do well.

APPLE IPHONE 13 (2021)

Apple released its iPhone 13 series on the go, with all four models going on sale at the same time, despite a worldwide processor shortage. There are still four models in the series—a 5.4-inch small, two 6.1-inch iPhones, and one 6.7-inch Pro Max—and they all follow the same formula.

The iPhone 13 differs from the iPhone 12 in that it has a new camera housing, a new CPU, and improvements to the display's notch.

APPLE IPHONE 13 MINI (2021)

The iPhone 13 mini, like the iPhone 12 mini, continued the trend and provided a tiny iPhone with equivalent power to the bigger iPhone 13. Given that there won't be an iPhone 14 small in 2022, it may be the final.

APPLE IPHONE 13 PRO (2021)

The iPhone 13 Pro, Apple's flagship device, had a new CPU and several improvements, such as a quicker refresh rate on the screen and a smaller "notch" at the top of the display. For those eager to flaunt the new model, there are a few new colors as well.

APPLE IPHONE 13 PRO MAX (2021)

Since the Apple iPhone 13 Pro Max and the iPhone 13 Pro have the same configuration, unlike prior models, the actual difference here is screen size rather than other factors like the camera. Many people who had previously believed they had to get the bigger model in order to access all of the new capabilities of the iPhone were pleased with that strategy.

The bigger battery in the iPhone 2021 series is without a doubt the biggest improvement. All four iPhones come with larger batteries than their predecessors, with the mini's battery capacity increasing by 10% and the Pro Max's battery capacity increasing by around 20%. This has led to a considerable increase in battery life, with the iPhone 13 Pro Max holding the record for the longest-lasting iPhone ever.

On the camera front, Apple unveiled Photographic Styles, a means to manage the color and contrast of all of your images. Unlike filters, Photographic Styles let you design a unique appearance for each shot you take and have it applied automatically. Additionally, all four iPhones include Theatrical Mode, which captures 1080p video with a backdrop that is blurry and an automated focus rack for a look that is more cinematic. The general impression in the community is that Cinematic Mode is interesting, but has some flaws and isn't completely finished, but it may still be enjoyable to use.

Additionally, the iPhone 13 series had brighter displays for better outdoor usage and a 20% reduced version of the iconic notch. The phones literally became physically bulkier and heavier with the larger batteries.

IPHONE SE (2022) MARCH 2022

Apple was prepared to debut a new member of the family, the iPhone SE, two years after the introduction of the second generation SE (2022).

Before its release, there wasn't much anticipation or enthusiasm for this iPhone, and it only represented a minor upgrade. The SE (2022) has the same form factor as the preceding iPhone SE, and because that design originated with the iPhone 6, it was clearly dated when it was first released. The tiny for the time 4.7-inch screen, which also uses the more antiquated LCD technology and has a lower than average resolution, was the issue that the majority of reviewers brought out.

On the other hand, Apple gave the SE (2022) an Apple A15 processor. In 2022, this was not just the fastest CPU available in a low-cost phone, but it was also the fastest processor available in any phone. Teardowns also showed an unexpected 10% increase in battery capacity, which was surprising given the identical form factor. Although the modem did not support mmWave, it did have the considerably more crucial C band frequencies for extensive coverage throughout the US, making it the first low-cost iPhone with 5G connection.

Finally, the beginning price for all of that was a little bit more, $430.

APPLE IPHONE 14 (2022)

The iPhone 14 is maybe the smallest update we've seen in a while since it maintains the same design and basic technology, adding just the new Crash Detection function and removing the actual SIM card port in the US.

The Apple iPhone 14 is conservative in both style and hardware, which makes the iPhone 14 Pro's offer more alluring this year than it could have been in years past.

Of course, there are some improvements. You get a better front camera, a better primary camera, and a little larger battery than the iPhone 13. Although there hasn't been much of a performance improvement and the design has remained the same, it is still a fantastic design.

APPLE IPHONE 14 PLUS

With a bigger display than the original model, Apple's iPhone 14 Plus represents a new approach for the company. With a 6.7-inch screen, it has the same basic specifications as the iPhone 14. Apple discontinued the "small" from their iPhone quartet and introduced the huge, 6.7" iPhone 14 Plus in its stead.

It's thrilling that the Apple iPhone 14 Plus introduces a new form factor for the basic iPhone. The iPhone 14 Plus is wonderful for those who want a larger screen but don't want to pay hefty Pro costs, but we at Pocket-lint adored the iPhone small and are a little bummed not to see one this year.

APPLE IPHONE 14 PRO

The iPhone 14 Pro has the same appearance as the 13 Pro, but it is more powerful thanks to the A16 Bionic processor, a new 48-megapixel primary camera, and a new Dynamic Island notch.

Of the four new iPhone 14 models, the Apple iPhone 14 Pro is without a doubt the most intriguing. It has a new look, as well as camera, CPU, and display advancements. The battery is also said to last longer, and Dynamic Island is awesome, so we won't ever complain about battery improvements.

APPLE IPHONE 14 PRO MAX

Everything about the iPhone 14 Pro is carried over to the bigger, 6.7-inch display of the iPhone 14 Pro Max. The new cameras, all the power, and the Dynamic Island experience are yours as a result.

The Apple iPhone 14 Pro Max is large in both shape and personality. Along with speed improvements, battery improvements, camera upgrades, the Always-On Display, and the Dynamic Island feature, there is a new design that is truly attractive, particularly in this bigger version.

This smartphone is great for people who prefer a bigger screen or better battery life, even though it is rather hefty, pricey, and has all the same functionality as its smaller sister.

CHAPTER TWO
THE IPHONE 14 AND IPHONE 14 PLUS

The iPhone 14 and iPhone 14 Plus are Apple's latest "budget" flagship iPhones, with pricing starting at $799. They will be sold alongside the more expensive iPhone 14 Pro and iPhone 14 Pro Max.

Apple eliminated the "small" iPhone in 2022, and the iPhone 14 models are available in 6.1- and 6.7-inch sizes. The term "iPhone 14 Plus," which alludes to the future 6.7-inch iPhone 14, pays homage to previous models such as the iPhone 8 and 8 Plus. Apple's iPhone 14 models include flat corners, an aerospace-grade aluminum shell, and a glass back that allows wireless charging, just like the iPhone 13 models.

The iPhone 14 variants are available in blue, purple, midnight, starlight, and (PRODUCT)RED. Unlike the iPhone 14 Pro, the TrueDepth camera is situated in a notch at the top of the display on the iPhone 14 and 14 Plus. The iPhone 14 and iPhone 14 Plus, like the iPhone 13 models, have Super Retina XDR OLED displays with 1200 nits of peak brightness, a contrast ratio of 2,000,000:1, Dolby Vision, and True Tone compatibility for adapting the display's white balance to ambient light. The standard iPhone 14 models do not incorporate ProMotion display technology; only the Pro models do.

The IP68 water resistance of the iPhone 14 versions allows them to survive a depth of 6 meters for up to 30 minutes. The iPhone's display is shielded from shocks and accidents by a durable Ceramic Shield front casing.

Apple is using the same A15 processor as in the iPhone 13 Pro models for the iPhone 14 models, but there has been an internal design change that allows for better thermal performance. The A15 processor is equipped with a 16-core Neural Engine, a 5-core GPU, and a 6-core CPU.

There is a 12-megapixel Wide camera with sensor-shift optical image stabilization, a larger sensor, an f/1.5 aperture, and larger 1.9 m pixels for better low-light performance. While there were no changes to the Ultra Wide lens, Apple did integrate a new TrueDepth camera with an enhanced /1.9 aperture. The Photonic Engine also improves low-light photography on the iPhone 14 cameras.

Using a combination of hardware and software, the Photonic Engine improves low-light performance by up to 2x on the Ultra-Wide camera, 2x on the TrueDepth camera, and 2.5x on the Wide camera. The Photonic Engine enhances Deep Fusion by adding incredible detail, preserving fine textures, and giving richer color.

There is a video action mode that provides slick-looking footage with enhanced picture stabilization. Even when video is captured during an incident, action mode may adapt to significant shaking, motions, and vibrations. The cinematic mode has been improved and is now available in 4K resolution at 30 and 24 frames per second. For more consistent illumination, the True Tone flash has higher consistency and is 10% brighter.

The dual-core accelerometer in the iPhone 14 models can measure up to 256Gs and drives the Crash Detection feature, which may instantly inform emergency personnel if you're in a catastrophic accident and can't reach your iPhone. Apple also uses additional sensors, such as the barometer, GPS, and microphone, to provide this capability. As a result, all of these sensors are capable of detecting variations in cabin pressure as well as abrupt changes in speed.

Furthermore, Apple updated the iPhone 14 to include Emergency SOS via satellite, a function that allows users to communicate in an emergency without WiFi or cellular coverage by connecting directly to satellites. Emergency SOS via satellite may be used to text emergency personnel in open areas with little trees. It is free for two years and works throughout the United States and Canada. When trekking or camping in remote areas, a satellite link can be used to broadcast your whereabouts to friends and family via Find My.

The iPhone 14 variants, like the iPhone 13 models, use a new Qualcomm X65 modem and support 5G connectivity (sub-6GHz and mmWave in the US). Because carriers now use eSIM, American iPhone models do longer include physical SIM cards.

The battery life of the iPhone 14 has increased, lasting an hour longer than on the iPhone 13. Because of its larger dimensions, the iPhone 14 Plus has a longer battery life than the iPhone 14. The iPhone 14 can last up to 20 hours when viewing video, while the iPhone 14 Plus can last up to 26 hours.

The iPhone 14 models have storage capacities of 128GB, 256GB, and 512GB. With a 20W or greater charger, fast charging via Lightning is available, but MagSafe charging up to 15W is still supported. WiFi 6 and Bluetooth 5.3 are also supported on the iPhone 14 and 14 Plus.

PRICING AND AVAILABILITY

The price of the iPhone 14 begins at $799 for 128GB of storage, while the price of the iPhone 14 Plus is $899 for the same amount of storage. A greater cost is charged for more storage. Pre-orders started on September 9 and the iPhone 14 launched on September 16; pre-orders started on September 9. On Friday, October 7, the bigger iPhone 14 Plus will be on sale and is now available for presale.

DESIGN

With the exception of new colors, the iPhone 14's design is virtually identical to that of the iPhone 13. It has the same flat sides and squared-off edges as last year's iPhones.

The iPhone 14 has a 6.1-inch screen, while the iPhone 14 Plus, which has a larger screen, is the same size as the iPhone 14 Pro Max at 6.7 inches.

The iPhone 14 variations have an all-glass front and a multicolored glass back, both wrapped in an aerospace-grade aluminum frame that matches their hues. The TrueDepth camera, speaker, and microphone are housed in a notch at the top of the display, as they have been on Apple's more affordable handsets. The iPhone 14 Pro and Pro Max have a "Dynamic Island" instead of a notch.

On the right is a power button, on the left are volume controls and a silent switch, and on the top and sides are antenna bands. A 5G mmWave antenna is situated below the power button, however it is only accessible on iPhone 14 models sold in the United States.

The bottom of the iPhone 14 and iPhone 14 Plus includes a Lightning connector for charging, speaker holes, and microphones. Since switching to eSIM, Apple has removed the SIM card slot from its products in the US.

The rear-mounted camera bump on iPhones still has a microphone, a True Tone flash, and a diagonal lens layout.

Except for the thickness, the iPhone 14 is identical to the iPhone 13. It is 5.78 inches long (146.7mm), 2.82 inches wide (71.5mm), and 0.31 inches deep (7.80mm). The majority of individuals will not notice the 0.15 mm difference. The iPhone 14 weighs 6.07 ounces less than the iPhone 13. (172 grams).

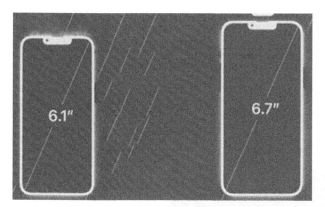

NO IPHONE 14 MINI

Apple marketed a 5.4-inch iPhone 13 small along with a 6.1-inch iPhone 13 last year, but there isn't a 5.4-inch iPhone this year. According to expert predictions, the iPhone 12 mini and iPhone 13 mini did not sell well, prompting Apple to stop making that specific model.

Apple currently offers the iPhone in two sizes, 6.1 and 6.7 inches, which correspond to the sizes of the iPhone 14 Pro and iPhone 14 Pro Max. Because the 6.7-inch iPhone was previously only available in "Pro" variants, this is the first time Apple has released a larger version of a standard iPhone.

COLORS

Since a few years ago, Apple has offered its regular iPhone lineup in a variety of vibrant hues. Blue, Purple, Midnight, Starlight, and red are the available colors for the iPhone 14 models.

WATER AND DUST RESISTANCE

The IP68 water and dust resistance rating applies to both the iPhone 14 and iPhone 14 Plus. The phones can endure a 30-minute submersion at a depth of six meters (19.7 ft).

The IP68 ratings of 6 dust resistance and 8 water resistance allow the iPhone 14 to withstand dirt, dust, and other particles... IP6x is the highest dust resistance standard available. Because of its IP68 rating, the iPhone 14 can endure splashes, rain, and accidental water contact; nevertheless, willful water exposure should be avoided.

According to Apple, dust and water resistance are not flawless and may deteriorate over time as a result of normal use. It is best to use caution while handling liquids because Apple's basic warranty does not cover liquid damage.

DISPLAY

There have been no display upgrades between the iPhone 14 devices and the iPhone 13 models from the prior generation. The iPhone 14 and iPhone 14 Plus's OLED Super Retina XDR display features a contrast ratio of 2,000,000:1 for darker blacks and brighter whites, as well as a peak brightness of up to 1200 nits for HDR photos, videos, TV shows, and movies. The standard iPhone 14 has a maximum brightness of 800 nits, which is 200 nits less than the 14 Pro varieties.

The iPhone 14 has a resolution of 2532 by 1170 with a pixel density of 460, while the iPhone 14 Plus has a resolution of 2778 by 1284 and a pixel density of 458.

Wide color support gives rich, true-to-life colors, and True Tone adapts the display's white balance to the surrounding illumination for a more comfortable viewing experience. A fingerprint-repellent oleophobic layer is also featured, as is support for Haptic Touch, which provides haptic feedback while operating the display.

The "Ceramic Shield" material protects the display from harm by fusing nano-ceramic crystals into the glass. Ceramic crystals were changed for the display, which was created in partnership with Corning, to maximize clarity while maintaining resilience. Thanks to a dual-ion exchange mechanism, Apple claims that Ceramic Shield is more resistant to scratches and daily wear and tear than any smartphone glass.

The iPhone 14 and 14 Plus do not support the ProMotion 120Hz refresh rate or the Always-In display technology present on the iPhone 14 Pro and Pro Max.

A15 BIONIC CHIP

This is the first time that Apple has not altered the A-series processor across the whole iPhone lineup; the iPhone 14 and 14 Plus share the same A15 chip as the iPhone 13 Pro and Pro Max. The A16 processor was not used in the iPhone 14 and 14 Plus, however it is in the iPhone 14 Pro and Pro Max.

Along with a 5-core GPU, the A15 contains a 6-core CPU with two performance cores and four efficiency cores. There were still gains compared to previous year as Apple only employed the A15's 5-core GPU version in the iPhone 13 Pro models and its 4-core GPU version in the iPhone 13 and 13 small.

While the CPU speed of the iPhone 13 and iPhone 14 is relatively identical, the GPU has seen some small improvements. Apple's internal architecture has also been updated to boost heat efficiency.

NEURAL ENGINE

Cinematic mode and Smart HDR 4 are powered by the A15 chip's 16-core Neural Engine, which can process up to 15.8 trillion operations per second.

RAM

The iPhone 14 and 14 Plus have 6GB RAM instead of the previous generation's 4GB.

STORAGE SPACE

The iPhone 14 and iPhone 14 Plus may be bought with up to 512GB of storage capacity, and all iPhone 14 models come standard with 128GB of storage. The iPhone 14 Pro variants have up to 1TB of storage capacity, which is half of this amount.

TRUEDEPTH CAMERA AND FACE ID

Face ID, the face recognition technology initially debuted in 2017, is still present on the iPhone 14 and iPhone 14 Plus. Apple did away with the notch on the iPhone 14 Pro models, but the display notch for the TrueDepth camera technology is still there on the ordinary iPhone 14 models.

Face ID is used across iOS to unlock the iPhone, grant access to passcode-protected third-party apps, validate in-app purchases, authenticate Apple Pay transactions, and more.

Face ID is enabled by a set of sensors and cameras. To build a 3D facial scan that maps the curves and planes of each face, a Dot Projector casts tens of thousands of invisible infrared dots onto the skin's surface. The scan is then read by an infrared camera.

The facial depth map is received by the A15 chip, which is then translated into a mathematical model that the iPhone uses for identification verification. Face ID is compatible with hats, beards, glasses, sunglasses, scarves, and other accessories that partially obscure the face in low light and darkness. Masks are now supported by Face ID in iOS 15.4.

FEATURES OF THE FRONT-FACING CAMERA

The new 12-megapixel front-facing camera on Apple's iPhone 14 models has an f/1.9 aperture, which lets in more light for better selfies and FaceTime video calls than before. It also includes focussing, which improves image quality and makes low-light selfies twice as excellent.

Many of the same features as the rear cameras are supported by the front-facing camera, including Night mode for selfies, Smart HDR 4, Dolby Vision HDR recording, Deep Fusion, and the new Photonic Engine, as well as ProRes and the Cinematic mode for recording videos with depth-of-field changes similar to those seen in movies.

The front-facing camera can also capture 4K video, QuickTake video, Slo-mo video, Portrait mode, Portrait Lighting, and use the Photographic Styles function to apply edits selectively.

DUAL-LENS REAR CAMERA

Apple now refers to the 12-megapixel Wide lens on the iPhone 14 and 14 Plus as the "Main" lens. A 12-megapixel Ultra Wide lens is also included in the dual-lens camera system.

Apple has enhanced the Main camera with an f/1.5 aperture over the /1.6 aperture used in the iPhone 13 camera. The new lens lets in more light, and it also has a larger sensor, which improves performance in low light. Apple states that the new Main camera allows in 49% more light than the iPhone 13 camera.

Unlike the 14 Pro and Pro Max, there is no Telephoto lens, and the Ultra Wide camera has not been improved. The iPhone 14 and 14 Plus can only zoom out digitally 5 times and optically 2 times.

PHOTONIC ENGINE

The Photonic Engine, a new software innovation in the iPhone 14, extends on Deep Fusion by inserting it early in the image process to add more information, maintain delicate textures, and improve image color.

Apple claims that the Photonic Engine provides a "dramatic boost" in mid- to low-light picture performance for both the Main lens and the Ultra Wide lens. Your primary camera's low-light photographs may be up to 2.5 times better than usual with Photonic Engine, and your ultra wide camera's low-light images may be up to 2 times better.

VIDEO CAPABILITIES

The iPhone 14 and 14 Plus are capable of recording 4K video at 24, 25, 30, and 60 frames per second. Recording modes include time-lapse, slo-mo, ProRes, and HDR.

Apple introduced a new Action mode to make video that looks smooth and ignores motion, tremors, and vibrations. Up to 2.8K at 60 frames per second can be recorded in action mode.

OTHER CAMERA FEATURES

Smart HDR 4 - Up to four persons may be recognized in a scenario, and each person's contrast, lighting, and even skin tone is optimized so they all appear their best.

Photographic Styles - Intelligent, moveable filters that may intensify or muffle colors while maintaining skin tone. Styles, as opposed to filters, which are applied universally to a picture, are applied selectively. Photographic styles include Vibrant (brightens colors), Rich Contrast (darker shadows and deeper colours), Warm (accentuates golden undertones), and

Cool (accentuates blue undertones). Because Tone and Warmth may be modified for each style, you can obtain the exact look you want.

Night mode - combines many photographs taken over the course of a few seconds to enable shooting in very low illumination conditions. Night mode exposures can be made up to two times faster with the Main iPhone 14 camera.

Deep Fusion - brings out the richness and complexity in the picture and works best in medium to low lighting settings.

Portrait mode - blurs the backdrop of the shot while keeping the subject in focus.

Portrait Lighting - Photographs taken in Portrait mode have their lighting altered using the Natural, Studio, Contour, Stage, Stage Mono, and High-Key Mono effects.

True Tone flash - The built-in flash, known as True Tone flash, is designed to blend in with the surroundings so that when it is utilized, the photo's white balance is not affected. The flash in the iPhone 14 is brighter and offers more stable brightness.

Panorama - Captures panoramic shots up to 63 megapixels.

Burst mode - allows for the simultaneous collection of several photos, which is advantageous for high-action scenes.

BATTERY LIFE

The battery in the iPhone 14 is 3,279mAh, which is somewhat larger than the battery in the previous generation's iPhone 13. The iPhone 14 Plus has a battery capacity of 4,325mAh. The iPhone 14 can play video for up to 20 hours, stream video for up to 16 hours, and play music for up to 80 hours, which is an hour longer than the iPhone 13.

The maximum video playback time on the iPhone 14 Plus is 26 hours, the maximum streaming video playtime is 20 hours, and the maximum audio playtime is 100 hours.

Both devices have fast-charge capabilities and may be charged to 50% in 30 to 35 minutes when using a power adapter of 20W or higher.

5G NETWORK CONNECTIVITY

The iPhone 14 models have a Qualcomm Snapdragon X65 modem that supports 5G connections. iPhone users in the United States can connect to both mmWave and sub-6GHz networks, but only sub-6GHz networks in other countries.

Because they are short-range and may be blocked by buildings, trees, and other obstructions, mmWave networks are limited to larger cities and metropolitan regions, as well as events like concerts, airports, and other locations where a large number of people gather.

Sub-6GHz 5G is becoming more common and accessible in cities, suburbs, and rural areas across the United States and other countries. When using a 5G network, you'll frequently be using Sub-6GHz 5G, which is often faster than LTE when using the mid-band spectrum, but it's not always the lightning-fast 5G you expect.

5G BANDS

More than 20 5G bands are supported by the iPhone 14 models sold in the US.

IPHONE 14 USER GUIDE

LTE BANDS

When 5G is not available, the iPhone 14 models continue to provide Gigabit LTE with 4x4 MIMO for connecting to LTE networks.

EMERGENCY SOS VIA SATELLITE

In an emergency, all iPhone 14 models may connect to satellites if cellular and WiFi access are absent. Apple's new Emergency SOS via satellite function connects the iPhone's antennae to Globalstar satellites.

The functionality is meant to be used in open areas with few trees so that the iPhone can connect to a satellite in the sky and be accessible in emergency situations when WiFi or cellular towers are unavailable.

Because sending data via satellite can take minutes, Apple has developed a compressed messaging protocol, and the company has a walkthrough interface that teaches how to hold the iPhone to connect to a satellite. In an area devoid of trees, a brief notification to emergency personnel can be transmitted in as little as 15 seconds.

When Emergency SOS via satellite is enabled, it asks you a number of crucial questions, the answers to which are intended to deliver the vital information to emergency services as soon as possible. If text communication is available in your region, the responses you provide are either routed directly to emergency services or to relay centers manned by Apple-trained professionals who may contact emergency services on the user's behalf.

When you're camping or trekking off the grid, you may use satellite internet to update your Find My Location so your family and friends know where you are. This offers them assurance that you are safe and sound. In life-threatening situations, satellite communication is primarily meant to allow users to contact emergency services.

Emergency SOS via satellite is free for all iPhone 14 customers for the next two years, although Apple has not stated how much it will cost in the future.

CRASH DETECTION

The iPhone 14 versions include a gyroscope with a high dynamic range and a dual-core accelerometer capable of measuring G-forces of up to 256Gs. This new technology, when paired with other iPhone sensors, such as a barometer that can detect changes in cabin pressure caused by deployed airbags, enables the usage of a crash detection feature.

When the automobile's occupants are critically injured or unable to access their iPhone, the iPhone 14 and 14 Plus may detect a significant car collision and instantly phone emergency services. For accuracy reasons, Apple trained motion detection algorithms on more than a million hours of actual driving and collision record data.

The iPhone's GPS can detect changes in speed, and its microphone may detect loud noises, which are frequently associated with major car accidents. When an accident occurs, the iPhone and Apple Watch work together to call emergency authorities via Crash Detection, which is also available on the Apple Watch.

BLUETOOTH, WIFI, NFC, GPS, AND U1 ARE ALL AVAILABLE.

Bluetooth 5.3 and WiFi 6 (802.11ax) with 2x2 MIMO are standard on the iPhone 14 and iPhone 14 Plus. An U1 Ultra Wideband chip and an NFC chip with reader mode are available. Because of the U1's increased spatial awareness, the iPhone 14 models can correctly locate other U1-equipped Apple devices.

When looking for AirTags, for example, the U1 chip may be used for exact tracking. It is also used for directional AirDrop and interactions with the HomePod mini equipped with an U1 chip. In addition to GPS, the iPhone 14 models support GLONASS, Galileo, QZSS, and BeiDou location services.

MAGSAFE

The rear of the iPhone 14 models still has MagSafe magnets that can be used to attach to magnetic devices like the MagSafe charger. Apple and approved third parties' official MagSafe chargers can wirelessly charge iPhone 14 models at up to 15W.

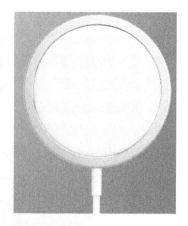

IPHONE 14 PRO AND PRO MAX

Along with the more costly iPhone 14 Pro and Pro Max, which represent a significant increase this year, Apple is also offering the iPhone 14 and 14 Plus.

IPHONE 14 PRO VARIANTS ARE PRESENT.

The iPhone 14 Pro and Pro Max are superior to the iPhone 14 models thanks to its 48-megapixel main camera, quicker and more effective A16 processor manufactured using a new 4-nanometer process, ProMotion display technology, and Always-On display.

FUTURE IPHONE RUMORS

The release of the iPhone 16 Pro in 2024 may herald the arrival of Apple's first all-screen smartphone. According to Apple analyst Ming-Chi Kuo, the phone might be the first to have both an under-screen front camera and Face ID, while according to display expert Ross Young, the iPhone 18 will first introduce under-display Face ID.

According to display researcher Ross Young, 2023 iPhones will likely have the same dual-hole cutout as the iPhone 14 Pro versions from this year. The same design with a circular cutout and pill-shaped hole will be used by the 2023 versions, and the feature won't be exclusive to the Pro models. It may be possible to reduce the size of the pill and hole design, but it will be a few years before Apple is ready for complete under-display Face ID.

USB-C

According to Bloomberg and Apple analyst Ming-Chi Kuo, in 2023 Apple will decide against using a Lightning connector for the iPhone in favor of a USB-C port. As a result, all iPhone 15 models will charge via USB-C instead of Lightning by default.

iPhone 15 Will Introduce USB C Port Instead of Lightning in 2023 feature

Although it has been reported by several sources that Apple would continue with Lightning, new laws being passed by the European Union would compel Apple to use USB-C on all of its products sold in Europe. If this law is passed, Apple would have to either modify all of its gadgets, which appears more probable, or sell special USB-C versions to Europe.

IN-HOUSE MODEM CHIPS

Similar to the Apple silicon and A-series processors, Apple is attempting to develop its own in-house built modem chips, which will enable the business to lessen its dependency on modem chip providers. Apple has been developing the modem for a number of years, and after purchasing the bulk of Intel's smartphone modem division in 2019, development intensified.

Apple might switch to its own 5G modems as early as 2023, according to analyst Ming-Chi Kuo. Apple won't need Qualcomm once it releases its own modem designs. The "earliest" timeframe is 2023, while other speculations claim that Apple chip supplier TSMC will be prepared to produce Apple's modem chips as early as 2023.

PERISCOPE LENS

The iPhone models from 2023 may include a periscope telephoto lens, which would enable more magnification without adding to the overall thickness of the phone.

UNDER-DISPLAY FACE ID

According to several speculations, Apple is attempting to eliminate the notch by putting the Face ID technology under the display. Initially, it was believed that Apple might use under-display Face ID technology as early as 2022, but display expert Ross Young predicts that it won't happen until 2023 or 2024.

A FOLDABLE IPHONE

Based on multiple reports and patents the firm has filed, as well as competition from businesses like Microsoft and Samsung, who have created foldable cellphones, Apple may eventually release a foldable iPhone.

CHAPTER THREE
THE IPHONE 14 PRO AND IPHONE 14 PRO MAX

The iPhone 14 Pro and iPhone 14 Pro Max, which were unveiled on September 7 and are being marketed alongside the more affordable iPhone 14 and iPhone 14 Plus, are Apple's latest premium flagship smartphones. The iPhone 14 Pro variants include many more features than the standard iPhone 14 models, including faster A16 processors, sharper displays, and advances in camera technology.

The flat edges, stainless steel enclosure, textured matte glass back, IP68 water resistance, and Ceramic Shield-protected display of the 6.1 and 6.7-inch iPhone 14 Pro models are similar to those of the iPhone 13 Pro models. However, the camera bumps are larger to accommodate new lenses, and the display has also changed.

The TrueDepth camera system is now housed inside of what Apple refers to as the Dynamic Island, a pill-shaped recess at the front of the display that holds camera technology and may alter in size and form based on what is shown on the screen.

According to Apple, the Dynamic Island aims to blur the distinction between software and hardware. It may adjust in real time, displaying notifications, alerts, and activities in the spot where the notch formerly was. Although there is still hardware there, the TrueDepth camera takes up less room, and the Dynamic Island uses the screen more effectively to blend it into the backdrop. The Dynamic Island may display purchase confirmations with Face ID, directions with Maps, phone calls, music information, Live Activities like sports scores, timers, and more. It can also be accessed by tapping and holding the screen.

A first for an iPhone, the new Super Retina XDR display on the iPhone 14 Pro and Pro Max supports an Always-On display thanks to enhanced ProMotion technology. A new 1Hz to 120Hz refresh rate and better power-efficient technologies enable the Always-On display. When the Always-On display is turned on, the background is darkened and the clock, widgets, and Live Activities are always visible at a glance.

The Super Retina XDR display on the iPhone 14 Pro, which is comparable to the Pro Display XDR in terms of peak HDR brightness at up to 2000 nits and Always-On technology, delivers greater peak HDR brightness. It has the greatest smartphone outdoor peak brightness and, in strong sunlight, is twice as bright as the iPhone 13 Pro.

The iPhone 14 Pro versions include an improved A16 processor, which is the first time Apple has utilized distinct chips for the iPhone 14 and the iPhone 14 Pro. The A16 is the first chip created using the new 4-nanometer manufacturing process, which improves speed and efficiency. The A16 boasts an upgraded 16-core Neural Engine that executes 17 trillion operations per second, a 6-core CPU that is quicker, and a 5-core GPU that is faster and has 50% higher memory bandwidth.

Apple included a 48-megapixel Wide camera with a quad-pixel sensor that adjusts to the image being shot, resulting in a 12-megapixel image with the same level of clarity as a 48-megapixel image. For even improved low-light performance, every four pixels in the majority of images are blended into one huge quad pixel. A 2x Telephoto mode, which employs the center 12 megapixels of the 48-megapixel lens to take full-resolution pictures without using digital zoom, is also available. Additionally enhanced, this adds to the current 3x zoom made possible by the specialized Telephoto lens.

The latest iPhones can capture stunning, expert images since the iPhone 14 Pro versions can shoot at the full 48MP with ProRAW. A new 12 megapixel Ultra Wide camera boasts bigger pixels for crisper shots with more information and a better macro mode, and second-generation sensor-shift optical image stabilization eliminates tremor and blur.

The TrueDepth front-facing camera now supports autofocus and has an enhanced f/1.9 aperture for better low-light selfies and movies. An array of nine LEDs that alter their pattern according to the lens focal length has been added to the new Adaptive True Tone flash. Night mode, Smart HDR 4, Portrait Mode, and Photographic Styles are still accessible features.

The Photonic Engine in the iPhone 14 Pro models is powered by Apple's A16 processor and enhances low-light performance across all of the lenses. With the Photonic Engine, the front-facing, wide, and telephoto TrueDepth cameras provide 2x improved low-light performance, while the ultra wide lens enables 3x better low-light performance. The Photonic Engine improves on Deep Fusion by adding amazing detail, maintaining fine textures, and delivering better color.

There is an Action Mode for video that offers more effective picture stabilization and smooth-looking footage than before. Even when video is recorded in the midst of an event, Action Mode may adapt to strong tremors, movements, and vibrations. The upgraded Cinematic Mode is now available in 4K at 30 fps and 4K at 24 fps.

The dual-core accelerometer in the iPhone 14 Pro versions can measure up to 256Gs. It also powers the Crash Detection function, which may call emergency services if you're in a catastrophic accident and are unable to reach your iPhone. In order to enable this function, Apple also makes use of additional sensors. As a result, the barometer can detect variations in cabin pressure, the GPS can monitor changes in speed, and the microphone may pick up the sounds of automobile crashes.

Additionally, Apple updated the iPhone 14 with Emergency SOS via satellite, a feature that enables users to communicate in an emergency without WiFi or cellular service by connecting directly to satellites. In open regions with few trees, emergency SOS via satellite may be used to text emergency personnel. It operates in the United States and Canada and is free for two years. When you're out hiking or camping in distant places, satellite connection may be utilized to broadcast your whereabouts with friends and family via Find My.

The new Qualcomm X65 modem is used in the iPhone 14 Pro variants, which allow 5G connection (sub-6GHz and mmWave in the US). Since carriers now employ eSIM, American iPhone models do not come with physical SIM cards.

With the iPhone 14 Pro and Pro Max, battery life has increased and now lasts an hour longer than the iPhone 13 Pro models. When viewing videos, the iPhone 14 Pro can last up to 23 hours, while the Pro Max can last up to 29 hours.

The iPhone 14 Pro comes with storage options of 128GB, 256GB, 512GB, and 1TB. With a 20W or greater charger, fast charging via Lightning is available, but MagSafe charging up to 15W is still supported. WiFi 6 and Bluetooth 5.3 are also supported on the iPhone 14 Pro and 14 Pro Max.

Deep purple, silver, gold, and space black are the available colors for the iPhone 14 Pro versions, with prices starting at $999. Pre-orders started on September 9 and the product was launched on September 16 after that.

PRICING AND AVAILABILITY

The price of the iPhone 14 Pro begins at $999 for 128GB of storage, while the price of the iPhone 14 Pro Max is $1,099 for the same amount of storage. A greater cost is charged for more storage. Pre-orders started on September 9 and the product was launched on September 16 after that.

DESIGN

The iPhone 14 Pro models have the same flat-sided, squared-off design as the iPhone 13 Pro models, which dates back to the iPhone 4.

The iPhone 14 Pro and Pro Max share virtually all of the same chassis components as the iPhone 13 Pro models, including an all-glass front and a textured matte glass back that is sandwiched between a frame constructed of surgical-grade stainless steel. For a sleek appearance, the rear glass's hue matches the color of the stainless steel frame.

For the iPhone 14 Pro versions, Apple eliminated the notch and replaced it with the "Dynamic Island," which we'll go into more depth about below. The TrueDepth camera system is housed in both a pill-shaped cutout and a circular cutout that make up the Dynamic Island's design. Apple combines the circular cutout and pill-shaped cutout into a single, bigger pill that is 30% smaller than the notch.

Along with a Lightning connector for charging, the phone has antenna bands on the top and sides. The left side of the device no longer has a SIM slot in the United States, however in certain other nations, the actual SIM hardware is still there. The front of the gadget is shielded by a robust Ceramic Shield display. According to Apple, the Ceramic Shield is created by fusing nano-ceramic crystals into glass and then toughening and enhancing it. The Ceramic Shield is designed to provide improved defense against scuffs and normal wear and tear.

The iPhone 14 Pro versions include a square camera bump on the rear, which is bigger than it was last year to accommodate new camera technology. There is a triple lens camera configuration, and each lens is bigger than the others.

SIZES

While the iPhone 14 Pro Max has a screen size of 6.7 inches, the iPhone 14 Pro is a 6.1-inch smartphone. It is Apple's biggest iPhone, matching the iPhone 14 Plus in size. The iPhone 14 Pro and Pro Max have somewhat different sizes when compared to the iPhone 13 Pro variants. The iPhone 14 Pro Max is somewhat more slender than the iPhone 13 Pro, which is slightly taller. Both variants are somewhat heavier and thicker.

The dimensions of the iPhone 14 Pro are 5.81 inches (147.5mm) high, 2.81 inches (71.5mm) broad, and 0.31 inches thick (7.85mm). The iPhone 14 Pro Max has dimensions of 6.33 inches (160.7) in height, 3.05 inches (77.6mm) in width, and 0.31 inches (0.3 cm) in thickness (7.85mm).

The Pro Max weighs 8.47 ounces (240 grams), whereas the Apple iPhone 14 Pro weighs 7.27 ounces (206 grams).

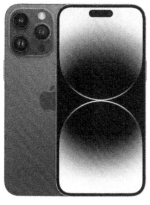

COLORS

The colors for the iPhone 14 Pro and Pro Max are Deep Purple, Space Black, Silver, and Gold. This year's hues are Space Black and Deep Purple, with Deep Purple replacing the Sierra Blue shade that was offered with the iPhone 13 Pro models.

WATER RESISTANCE

The IP68 water resistance certification applies to the iPhone 14 Pro and Pro Max. Similar to the iPhone 13 Pro variants, they can survive fluids up to six meters deep (19.7 feet) for up to 30 minutes.

IP68 is broken down into two parts: 6 for dust resistance (meaning the iPhone 13 Pro can withstand dirt, dust, and other particles), and 8 for water resistance. The maximum dust resistance grade available is IP6x. The iPhone 14 Pro can withstand splashes, rain, and accidental water exposure thanks to its IP68 water resistance classification, although it should be avoided if at all possible.

According to Apple, dust and water resistance are not infallible states and may degrade over time as a consequence of ordinary use. Since liquid damage is not covered by Apple's warranty, it is essential to exercise care while subjecting the iPhone 14 Pro models to liquids.

DISPLAY

The improved OLED Super Retina XDR display on the iPhone 14 Pro and Pro Max is flexible and extends into the body of each handset. A 2,000,000:1 contrast ratio enables for blacker blacks and brighter whites, and the display has a peak brightness of up to 2000 nits when outside, making it easier to view in direct sunlight.

The display is substantially brighter than the iPhone 13 Pro display and supports a peak brightness of 1600 nits in HDR, which is higher than the typical maximum brightness of 1000 nits.

With 460 pixels per inch, the 6.1-inch iPhone 14 Pro has a resolution of 2556 x 1179, while the 6.7-inch iPhone 14 Pro Max has a resolution of 2796 x 1290.

Wide color support creates vibrant, true-to-life colors, and True Tone adjusts the display's white balance to the ambient illumination of the space it is in for a viewing experience that is designed to be easier on the eyes and more like reading on paper. A fingerprint-repellent oleophobic layer is also featured, as is support for Haptic Touch, which provides haptic feedback while operating the display.

DYNAMIC ISLAND

The iPhone 14 Pro and Pro Max versions do not have a notch, as Apple has abandoned a style it has utilized since the iPhone X in 2017. The proximity sensor has been repositioned underneath the display, and smaller cuts have been used for the camera, the dot projector, and other components. This is how Apple has updated the TrueDepth camera hardware.

The right side of the display has a tiny circular cutout for the camera, while the left side has a pill-shaped cutout for the TrueDepth system for Face ID. These two cutouts were combined into one pill-shaped cutout that Apple is referring to as the "Dynamic Island" utilizing the software.

Since Apple made the Dynamic Island a functional component of the interface, it is more than simply a cutout. It transforms and changes shape to fit what is shown on the iPhone's screen and serves as a type of information hub that is front and center.

Apple uses the Dynamic Island to provide quick access to information about the programs and services you're using on your iPhone, blending it into the display. For example, Dynamic Island extends into a square form to match the Face ID confirmation screen when you make an Apple Pay payment, and it expands to be bigger while you're on a call so you can have phone functions right at your fingertips.

It may show the connection status for the AirPods, Apple Music, timers, and Maps directions. For monitoring sports results, Uber journeys, and other things straight from the top of the iPhone's display without leaving the app you're in, it connects with Live Activities in iOS 16.

If you want to view what's on your display, you can fold it down. If you touch it, however, it will open up so that you may interact with the material. There is a feature that enables it to break out into two distinct cutouts that each display a different piece of information, and third-party applications may also add support for Dynamic Island.

PROMOTION

The iPhone 14 Pro and Pro Max, like the iPhone 13 Pro models, have enhanced low-power display backlighting that supports ProMotion refresh rates. The adaptive refresh rates on this year's iPhones vary from 1Hz to 120Hz, compared to 10Hz to 120Hz on last year's models.

The refresh rate of the display varies depending on what is displayed on the screen. When reading a static page, the iPhone will use a lower refresh rate, while playing games, watching movies, or scrolling through information, it will use a higher 120Hz refresh rate. This results in a more fluid and responsive experience.

ALWAYS-ON TECHNOLOGY

A first for an iPhone, the new 1Hz refresh rate choice allows the Always-On display functionality. Although always-on display technology has been available on the Apple Watch for a while, this is the first time Apple has been able to make it function on an iPhone without using up too much power.

The wallpaper is muted to give it a modest appearance while the time and widgets remain live on the Lock Screen with the Always-On display. When the display is not in use, the status bar, flashlight, and camera buttons are concealed.

The screen does entirely shut down when the iPhone is placed face down or in a pocket to save battery life.

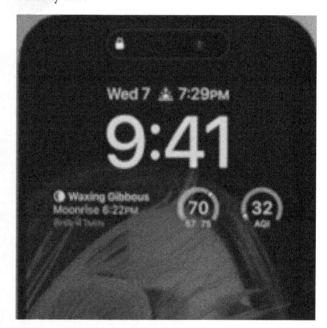

A16 CHIP

For the iPhone 14 Pro and Pro Max, Apple created an A16 Bionic processor, the company's first to be produced using a brand-new 4-nanometer manufacturing process. It contains four high-efficiency cores and two high-performance cores, and thus yet, it has shown just minor advancements over the A15 processor found in the iPhone 13 Pro.

The A16 scored 1887 on the single-core test and 5455 on the multi-core test, whereas the A15 scored 1707 on the single-core test and 4659 on the multi-core test. That is an improvement of 10.5, respectively, 17.1%.

An accelerated 5-core GPU with 50% higher memory bandwidth is included in the A16 Bionic.

NEURAL ENGINE AND ISP

The A16 has an upgraded 16-core Neural Engine that can do around 17 trillion operations per second.

Additionally, the iPhone 14 Pro and Pro Max's image signal processor, which does up to four trillion operations in each picture, is part of the A16 chip.

RAM

The iPhone 14 Pro and Pro Max both have 6GB of RAM, which is the same amount as the iPhone 13 Pro versions. When compared to the iPhone 13 Pro models from the previous generation, which used LPDDR4X memory, the performance of the iPhone 14 Pro and Pro Max is improved thanks to the faster LPDDR5 memory.

STORAGE SPACE

The iPhone 14 Pro and Pro Max's entry-level models come with 128GB of storage space, however there are other models with 256GB, 512GB, and 1TB of storage space.

TRUEDEPTH CAMERA AND FACE ID

Face ID, the face recognition technology that was initially released in 2017, is compatible with the iPhone 14 Pro models. The TrueDepth camera system, which is placed in the pill-shaped cutout on the front of the display, is where the Face ID components reside.

Face ID may be used to unlock an iPhone, verify transactions, reset passwords, and perform other functions. Face ID is enabled by a system of sensors and cameras; a Dot Projector beams tens of thousands of invisible infrared dots on the skin's surface to build a 3D facial scan that stores the contours and planes of each face.

The depth map is received by the A16 chip, which turns it into a mathematical model that the iPhone uses for identification verification. Face ID works in low light and the dark with hats, beards, glasses, sunglasses, scarves, and, beginning with iOS 15, face masks.

FRONT FACING CAMERA FEATURES

You can shoot better selfie photos and movies in low light with a new front-facing TrueDepth camera with an f/1.9 aperture. For the first time, the front-facing camera has autofocus, allowing it to focus faster in low-light situations and snap group selfies from a greater distance.

TRIPLE-LENS REAR CAMERA

Wide Lens

The iPhone 14 Pro and Pro Max include an improved 48-megapixel Wide lens with quad-pixel picture stabilization and second-generation sensor-shift optical image stabilization. The lens has an f/1.78 aperture and a 24mm focal length.

The 48MP shot is effectively reduced to a 12MP photo with remarkable quality and low-light performance since the combined pixels allow for 4x more light. The quad-pixel sensor combines every four pixels into a single pixel that is comparable to 2.44 m in order to make file sizes acceptable.

The new 2x Telephoto option, made possible by the quad-pixel sensor, employs the center 12 megapixels of the sensor to capture full-resolution images without the need for digital zoom.

In ProRAW mode, the iPhone can take ProRAW pictures at their full 48-megapixel resolution for greater information, according to Apple, opening up new creative processes for professionals. The 48MP lens may be used by third-party camera applications to shoot full-resolution pictures.

Ultra Wide Lens

In order to produce crisper photographs with more information and to give more detail for macro shots, a new 12-megapixel Ultra Wide lens has bigger 1.4 m pixels.

The six-element lens has a 120 degree field of view, an f/2.2 aperture, and a 13mm focal length.

Telephoto Lens

A 3x optical zoom is available on the upgraded telephoto camera, which may be utilized in addition to the 2x zoom offered by the wide camera. The 77mm, six-element lens supports optical image stabilization and has an f/2.8 aperture.

PHOTONIC ENGINE

A new software tool called The Photonic Engine expands on Deep Fusion by using it early in the imaging process to add more detail, maintain fine textures, and enhance the color of the image.

For mid to low-light picture performance across all cameras, according to Apple, the Photonic Engine delivers a "huge jump." The Wide camera has up to a 2x improvement in low-light performance, while the Ultra Wide camera has a

3x improvement, the Telephoto camera has a 2x improvement, and the front-facing TrueDepth camera has a 2x improvement.

VIDEO CAPABILITIES

The iPhone 14 Pro and Pro Max can record 4K video at 24, 25, 30, or 60 frames per second. Recording modes include time-lapse, slo-mo, ProRes, and HDR. Keep in mind that capturing ProRes film in 4K requires a 256GB iPhone 14 Pro model.

Apple introduced a new Action Mode that is designed to provide smooth video while ignoring motion, rattling, and vibrations. In Action Mode, you can capture up to 2.8K at 60 frames per second.

OTHER CAMERA FEATURES

Adaptive True Tone flash - The flash has been modified with a nine-LED array that may change pattern depending on the focal length of the lens. The flash is up to two times brighter on Telephoto pictures **and three times brighter on Ultra Wide shots.**

Cinematic Mode - Cinematic mode is available in 4K at 30 frames per second and 4K at 24 frames per second.

Macro Photography - The Ultra Wide cameras on the Pro versions feature a 2 cm focus distance, making them ideal for macro photography. Slow motion and time-lapse macro films, as well as macro pictures, can be captured.

Smart HDR 4 - Up to four persons may be recognized in a scenario, and each person's contrast, lighting, and even skin tone is optimized so they all appear their best.

Photographic Styles - Photographic Styles are intelligent, customizable filters that can amplify or mute colors without changing skin tone. Styles are applied to an image selectively, as opposed to filters, which are applied to the entire image. Photographic styles include Vibrant (which enhances colors), Rich Contrast (which creates darker shadows and deeper hues), Warm (which emphasizes golden undertones), and Cool (accentuates blue undertones). Tone and warmth are adjustable for each design, allowing you to achieve the precise look you desire.

Night Mode - combines several images taken over the span of a few seconds to allow shooting in low-light circumstances. Night Mode is compatible with Portrait Mode and all three lenses thanks to the LiDAR scanner in the Pro versions.

Deep Fusion - brings out the richness and complexity in the picture and works best in medium to low lighting settings. On the iPhone 14 models, it functions even sooner throughout the picture capture process thanks to Photonic Engine.

Portrait Mode - blurs the backdrop of the shot while keeping the subject in focus.

Portrait Lighting - adjusts the lighting in Portrait Mode images using the Natural, Studio, Contour, Stage, Stage Mono, and High-Key Mono effects.

Panorama - up to 63 megapixels of panoramic images may be captured.

Burst Mode - allows for the simultaneous collection of several photos, which is advantageous for high-action scenes.

BATTERY LIFE

A 3,200mAh battery powers the iPhone 14 Pro, while a 4,232mAh battery powers the larger iPhone 14 Pro Max. Up to 23 hours of video playback, 20 hours of streaming video, and 75 hours of audio playing are all possible on the iPhone 14 Pro. Up to 95 hours of music may be played back and up to 29 hours of streaming video can be played back on the iPhone 14 Pro Max.

The iPhone 14 Pro versions have around an hour more battery life for watching videos compared to earlier iPhone models.

With a power adapter of 20W or above, both devices have rapid charging capabilities and can be charged up to 50% in 30 minutes.

5G CONNECTIVITY

The iPhone 14 Pro models have a 5G-capable Qualcomm Snapdragon X65 modem. iPhone users in the United States can connect to both mmWave and sub-6GHz networks, but only sub-6GHz networks in other countries.

Because they are short-range and may be blocked by buildings, trees, and other obstructions, mmWave networks are limited to larger cities and metropolitan regions, as well as events like concerts, airports, and other locations where a large number of people gather.

Sub-6GHz 5G is becoming more common and accessible in cities, suburbs, and rural areas across the United States and other countries. When using a 5G network, you'll frequently be using Sub-6GHz 5G, which is often faster than LTE when using mid-band spectrum, but it's not always the lightning-fast 5G you expect.

5G BANDS

More than 20 5G bands are supported by the iPhone 14 Pro versions sold in the US.

LTE BANDS

When 5G is not available, the iPhone 14 models continue to provide Gigabit LTE with 4x4 MIMO for connecting to LTE networks.

EMERGENCY SOS VIA SATELLITE

Models of the iPhone 14 may now connect to satellites in instances when cellular and WiFi connectivity are unavailable. Apple has created the new Emergency SOS via satellite function to link the iPhone's antennae to Globalstar satellites.

In emergency scenarios when you cannot connect to WiFi or a cellular tower, satellite communications are available. This capability is meant to be utilized in broad areas with few trees so the iPhone can connect to a satellite in the sky.

IPHONE 14 USER GUIDE

Because transferring data via satellite may take minutes, Apple has designed a compressed messaging protocol and a graphical interface for connecting the iPhone to a satellite. A brief communication to emergency responders may be sent in as little as 15 seconds in an area devoid of trees.

When triggered, Emergency SOS via satellite prompts you with a series of vital questions designed to give emergency services with the information they need as fast as possible. If text communication is available in your region, your responses are transmitted immediately to emergency services; otherwise, they are routed to relay centers staffed by Apple-trained professionals who may contact emergency services on the user's behalf.

When you're camping or trekking off the grid, you may use satellite internet to update your Find My Location so your family and friends know where you are. This offers them assurance that you are safe and sound. In life-threatening situations, satellite communication is primarily meant to allow users to contact emergency services.

Emergency SOS via satellite is free for all iPhone 14 customers for the next two years, although

Apple has not stated how much it will cost in the future.

CRASH DETECTION

This new technology enables the iPhone to detect collisions owing to a high dynamic range gyroscope, a dual-core accelerometer that can measure G-forces up to 256Gs, and other sensors including a barometer that can detect cabin pressure changes caused by deployed airbags.

When the passengers of the vehicle are gravely hurt or unable to access their iPhone, the iPhone 14 models may detect a significant automobile collision and instantly summon emergency services. For accuracy reasons, Apple trained motion detection algorithms on more than a million hours of actual driving and collision record data.

The iPhone's GPS can detect changes in speed, and its microphone may detect loud noises, which are frequently associated with major car accidents. When an accident occurs, the iPhone and Apple Watch work together to call emergency authorities via Crash Detection, which is also available on the Apple Watch.

BLUETOOTH, WIFI, NFC, GPS, AND U1

Bluetooth 5.3 and WiFi 6 (802.11ax) with 2x2 MIMO are standard on the iPhone 14 Pro and Pro Max. An U1 Ultra Wideband chip and an NFC chip with reader mode are available. Because of the U1's increased spatial awareness, the iPhone 14 models can correctly locate other U1-equipped Apple devices.

When finding AirTags, for instance, the U1 chip may be employed for precise tracking. Additionally, it is used for interactions with the HomePod mini equipped with an U1 chip and for directional AirDrop. Regarding GPS, the iPhone 14 versions support L1 and L5 frequencies for better positioning accuracy, as well as GPS, GLONASS, Galileo, QZSS, and BeiDou location services.

MAGSAFE

The rear of the iPhone 14 Pro versions still include MagSafe magnets, which can be used to attach the MagSafe charger and other magnetic accessories. The iPhone 14 Pro variants may be wirelessly charged at up to 15W using authorized Apple MagSafe chargers.

There is also Qi-based charging, although its maximum output is 7.5W.

7 BIG CHANGES

On September 7, 2022, Apple unveiled the iPhone 14 as well as new Plus, Pro, and Pro Max variants. There are some significant changes this time around, including new satellite connection capabilities on every model and a redesigned notch on the iPhone 14 Pro.

NO MORE SIM CARD SLOT (IN THE US)

With its choice to do away with features like optical discs, headphone connectors, and USB-A connections, Apple has already the made news. The business is at it again, leaving a real SIM card slot out of the US models of the iPhone 14. You must use an eSIM in place of a SIM card to connect the iPhone 14 or 14 Pro to a cellular network.

The abbreviation "eSIM" stands for "embedded SIM," and it enables carrier switching without removing a physical SIM card. Since the iPhone XS, the function has been available in several previous iterations of Apple's smartphone. It has never before been necessary for US consumers to utilize it.

By setting a choice in iOS Settings, carriers may be changed instantly in place of replacing a real SIM card. The iPhone 14 and iPhone 14 Pro both have storage capacities of six and eight, respectively. The iPhone 14 lineup's physical SIM card slot will continue to be supported on international variants (for now)

If you use a large carrier, the change shouldn't be too problematic, but anybody using a smaller, more affordable carrier should first confirm that they support an eSIM before making a purchase. The choice to eliminate the SIM card slot reduces the number of points of entry that need to be considered for water and dust protection, which should speed the adoption of eSIM by cellular operators throughout the globe.

NEW SATELLITE CONNECTIVITY

Emergency SOS via Satellite is a brand-new feature that will be included on both 2022 iPhone models. As the name implies, this enables the iPhone to make a satellite network connection in an emergency when it is needed and cellular service is not available. Beginning in November, the service will be accessible in the US and Canada.

Thanks to Apple's "relay centers," which forward text messages, satellite communication may communicate with SOS services even those that only take voice calls. Apple also enables non-emergency uses, such as informing loved ones that you are OK when away from a cellular network.

Apple has said that the service would be "free" for two years, but has not specified what the cost of the function will be beyond that time. Apple may add the service to its iCloud+ subscription tier, which currently includes services like Hide My Email and iCloud Private Relay when you purchase extra iCloud storage, or charge an additional cost.

NO IPHONE MINI FOR 2022

The iPhone 14 currently has a 6.1-inch screen as standard equipment, with a 6.7-inch iPhone 14 Plus model available as an option for customers who want a bigger screen. Sadly, it seems that Apple no longer caters to individuals who like smaller smartphones, since the 5.4-inch iPhone 13 mini will not get an upgrade in 2022.

You can still purchase the iPhone 13 small from last year on the Apple website, but at just $200 less expensive than the $799 iPhone 14, it could be difficult to convince customers given the lower battery life (a quoted three hours shorter than the iPhone 14).

It's too soon to conclude that Apple has discontinued the smaller iPhone models completely since it's unclear if the company will release a new iPhone mini in 2019. Apple may decide to switch between the Plus and small cycles, which may be a good idea given that most users don't update their iPhones yearly anyhow.

CRASH DETECTION IN ALL MODELS

With the release of the iPhone 14's upgraded "high dynamic range" gyro and new "high-g" accelerometer, Apple has included a function dubbed "crash detection" (and Pro). The iPhone can recognize when you've been in an automobile accident and offer to call emergency personnel for you, similar to Fall Detection on the Apple Watch.

The iPhone will continue to call for assistance and Siri will provide your location if you don't answer to the request.

The function is based on information acquired by onboard sensors, such as unexpected changes in speed and direction, loud noise levels, and variations in cabin pressure brought on by airbag deployment.

A BETTER NOTCH (IPHONE 14 PRO)

On the iPhone 14 Pro, Apple has redesigned the contentious notch and given it a ridiculous new name. The makeover, now referred to as the Dynamic Island, more closely resembles the pill-shaped cuts featured on competing Android handsets. It contains the FaceTime camera on the front, the ambient light sensor, and the sensors needed for FaceID to unlock the phone..

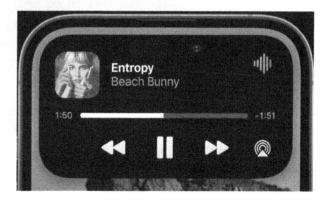

But Apple has gone a step further and effectively repurposed the unused area within iOS, making it useful once again. The sensor array now seamlessly integrates with turn-by-turn navigation, notifications, alarms, controls for Now Playing, and more. The Dynamic Island may be touched and held to display information, including playback settings and other details, along with a stylish animation.

Apple was able to reduce the TrueDepth camera system for Face ID by 31% while also conserving space by moving the proximity sensor beneath the display. Future updates to the basic iPhone models should have this capability.

BRIGHTER, ALWAYS-ON DISPLAY (IPHONE 14 PRO)

The iPhone 14 Pro has an always-on display to go along with Apple's new lock screen widgets, which are prominent in iOS 16. The function was initially seen in the iOS 16 beta, sparking speculation that Apple would release it for the iPhone 13 Pro. Although theoretically still viable, Apple has only said that the capability is available on the iPhone 14 Pro.

When the iPhone is put in a pocket or face down, the display detects this using the proximity sensor and turns down the display to save battery life. If the implementation of the function is anything like the Apple Watch's always-on display, there will be a toggle in Settings. However, Apple has not said if the feature can be completely turned off or not.

Additionally, the display can now achieve brightness levels of 1,600 nits in HDR images and movies (up from 1,200 nits on last year's model) or 2,000 nits in strong daylight, which is twice as bright as the iPhone 13 Pro.

IMPROVED CAMERA AND FLASH (IPHONE 14 PRO)

The camera system is likely the largest improvement for the iPhone 14 Pro (due to the A16 Bionic system-on-chip that runs the smartphone), even if it is less immediately obvious than features like Dynamic Island and the always-on display. The 12-megapixel sensor seen in the iPhone 13 Pro has been replaced with a huge 48-megapixel sensor in the primary iPhone camera.

A 65% bigger sensor, which allows in more light with each photo for even greater low-light performance, is what makes it feasible to increase resolution by four times. Along with the already available 3x telephoto and 0.5x ultra-wide, there is now a new 2x telephoto option with an effective focal length of 48mm. Utilizing the "middle 12 megapixels" of the 48-megapixel sensor, the 2x zoom is made feasible for 4K quality photographs without the need for digital zoom.

The iPhone 13 Pro's Cinematic mode has been improved with the ability to film in 4K HDR at

24 (or 30) frames per second instead of the previous 1080p. Action mode is a new feature that allows you to capture smooth handheld video without the need for a gimbal.

Last but not least, a revamped TrueTone flash with 9 LEDs has been introduced. Depending on the focal length and composition you choose, the flash may change its pattern and intensity.

Apple has promised a new "Advanced" dual camera system with greater low-light performance on the basic model this time around, even if the iPhone 14 Pro gets the most of the changes in this update.

INCLUDES IOS 16

On September 16, the new iPhone 14 will be available for $799, and on October 7, the iPhone 14 Plus will start at $899. On September 16, the iPhone 14 Pro (from at $999) and Pro Max (starting at $1099) will also be accessible.

CHAPTER FOUR
THE EASIEST WAY TO SET UP YOUR NEW IPHONE 14

Congratulations! Your new iPhone 14 has arrived (or you ordered it and are now impatiently expecting it). So, how do you activate your new iPhone? What can you anticipate during setup? The stages for a seamless transfer are listed below.

1. UPDATE YOUR CURRENT IPHONE

To make sure everything goes well during the data transfer, first and foremost upgrade your old iPhone to the most recent version (currently iOS 15.7 for most phones).

DO NOT DOWNLOAD iOS 16 TO AN OLD IPHONE.

Apple readily acknowledges that the first release of iPhone 16, which comes pre-loaded on iPhone 14, has received complaints about being "extremely" problematic, especially with regard to Messages and FaceTime. Therefore, the current version of iOS 16 on your old iPhone may be too problematic to successfully finish a data transfer to a new iPhone 14. However, if you have upgraded your outdated phone to iOS 16, don't be alarmed. The Editor-in-Chief of Techlicious, Suzanne Kantra, and many others who upgraded to iOS 16 had no trouble transferring their data to an iPhone 14.

Update your new iPhone 14 to iOS 16.0.1 if requested to do so. Do not be concerned if you are not prompted. At the conclusion of the setup procedure, you may upgrade to the most recent version of iOS 16.

2. BACK UP YOUR CURRENT IPHONE

Back up your existing iPhone after updating it to the newest version of iOS. For your new device's setup, you'll need a brand-new backup.

The material, information, and passwords on your current iPhone may be backed up to iCloud or your computer. Backing up to iCloud should take 5–10 minutes, whereas backing up to your PC would likely take twice to three times as long depending on how much data you have on your iPhone.

Go to Settings and tap on your Apple ID name or picture > iCloud > iCloud Backup > Back Up Now to backup your current iPhone (this should be the third item listed). A progress meter and the predicted amount of time are shown.

Connect your iPhone to your computer using a Lightning-to-USB connector to back up your current iPhone. Your iPhone will show up as a listed device or icon in Finder on a Mac, and in the upper left corner of the iTunes program window on a Windows computer.

When you double-click the iPhone icon in Finder on a Mac or click the iPhone symbol in iTunes on a Windows computer, a window similar to this one will appear:

You wish to manually backup your iPhone. Select "Back up all of your iPhone's data to this." PC or Mac. Encrypt your backup if you choose. Apple cautions you that if you decide not to encrypt your backup, "Without encryption, sensitive data won't be included in the backups. Your stored passwords, Health, and HomeKit data will be lost if you restore a device from a backup." You'll be requested to create a password for this encrypted backup if you haven't previously. You'll click "Back Up Now" on your iPhone after entering your PIN number.

You are prepared to configure your new iPhone 14 after you have completed backing up your soon-to-be-ex iPhone.

3. ACTIVATE YOUR NEW IPHONE

The first step is to switch on your new iPhone 14 (it should arrive fully charged, but you are welcome to connect it to power) and go through the location and language selection screens.

You may decide how to transfer data to your new iPhone, but to activate it and start the iPhone-to-iPhone transfer, pick Quick Start.

The old and new iPhones just need to be placed next to one another. You will see what appears to be a teal-colored cloud on your brand-new iPhone 14. On your previous iPhone, a camera window will then appear. Finally, you place the teal cloud on your new iPhone 14 over the old iPhone camera, and in less than five minutes, your new iPhone 14 is prepared for data transmission.

4. SWAP YOUR TELEPHONE NUMBER

You will be prompted to transfer your existing phone number to the new iPhone 14 when it is available, which we presume you wish to do. The traditional SIM card, which is the size of a pinky nail, is replaced with an electronic SIM, or eSIM, in the iPhone 14 model. Just choose Transfer from Another iPhone from the menu. It took me a few tries for some reason, but eventually my number was correctly transferred.

If you own a Mac, you may anticipate seeing a number of alerts informing you of FaceTime and Message updates as well as opportunities to enable WiFi calling.

5. TRANSFER YOUR DATA

After that, a request to transfer your data will appear. You have the option of transferring from your local computer backup or your iCloud backup. If you pick iCloud, you may start using your new iPhone in around 15 minutes since the majority of your data will begin downloading in the background after that. It can take many hours or longer to transfer data from your Mac or PC via a USB cord. Therefore, I advise selecting to transfer from iCloud.

You are then given the option to choose precisely what data you want to transfer before being asked whether you want to Make This Your New iPhone. If you want to make your new iPhone 14 a copy of your old iPhone, just select "Continue."

We found that the iCloud iPhone to iPhone 14 transfer took exactly 15 minutes as predicted. However, not everything has yet been transferred.

Your phone number and financial information must be exchanged after the contents of your old iPhone have been securely migrated via iCloud.

You must enter the security codes for any registered credit cards or Apple Cash accounts in order to add them to your new iPhone.

You'll be prompted to turn on the Emergency SOS feature on your new iPhone. You'll also be informed that the iPhone 14's new Emergency SOS via Satellite will be available in the fall.

Additionally, you'll be given the opportunity to generate a new FaceID with the following options: without the Covid mask, with the mask, and with or without glasses. You can also add "alternative" faces, such as those of a loved one, yourself sporting a different cap or hairstyle, etc. You can forego adding these for the time being and set them up after your iPhone 14 is fully prepared.

You'll be given the option to wipe your previous iPhone before being asked what you want to do with it. Don't delete your old iPhone if you use an authenticator app. If you still have access to your old iPhone, transferring your authenticator account to the new iPhone will be significantly simpler. I like to continue using my old phone (which is now only powered by WiFi) until I'm certain everything is working properly with my new iPhone.

Your iPhone 14 will shut down and restart after the transfer of your phone's contents is finished.

6. COMPLETE THE DOWNLOADS

Unbelievably, your iPhone 14 content download isn't complete if you use iCloud to transfer your info. When your iPhone 14 resumes, all the content you had stored on your old iPhone (apps, images, videos, music, etc.) is busy downloading while you set up FaceID and your financial information. Your new iPhone 14 was simply informed of the content on your previous iPhone during the 15-minute iCloud data transfer. These files are currently uploading to your new iPhone.

This data download procedure could take an hour or longer, depending on how much actual data you had stored on your old iPhone. You can view how much data has to be downloaded to your new iPhone by going to Settings. Although you wouldn't be able to use your new iPhone during this wired PC-to-iPhone 14 transfer, the combined iCloud transfer and download time is still quicker.

You can thus use your new gadget while this download is running. Nevertheless, we advise waiting.

7. UPDATE YOUR IPHONE 14

> **Restore in Progress**
> An estimated 8.7 GB will be downloaded to finish restoring your apps and data.

> **Important Software Update**
> An update is available to address an iMessage and FaceTime activation issue.
> Update Now

As previously mentioned, Apple promptly released iOS 16.0.1 as a fix for the observed issues with Messages and FaceTime in the initial release of iOS 16. Update to iOS 16.0.1 on your iPhone 14 now. Apple includes a reminder to do this in the primary Settings screen.

Even with all of this transferring, you may still need to manually enter the passwords for some apps, email accounts, and other internet accounts and subscriptions. You will also be invited to sync your new iPhone with your Apple Watch if you own one.

The name of your new iPhone 14 will be the same as your previous one because it was copied from your current iPhone. Go to Settings, then General > About > Name to change the name on your new iPhone.

Additionally, alter your email signatures by navigating to Settings > Mail > Signature if you reply to emails using the name of your iPhone.

You can now use your iPhone 14 and iOS 16 to your heart's content.

CHAPTER FIVE
HOW TO USE SIRI ON IPHONE 14

Numerous actions may be completed by Siri on your iPhone. Be it simple activities like making a reminder for you, setting a timer or an alarm, or complex ones like phoning someone, reading out a message, sending a message, and hanging up calls. As if that weren't enough, you can use Siri to intercom people or share material from a few applications with one of your contacts.

The list continues. On the iPhone 14 series, all you have to do is ask Siri to do the tasks for you. However, you must first set up Siri and turn on the "Hey Siri" function before you can start asking Siri to do tasks for you.

CONFIGURE SIRI TO YOUR PREFERENCE

When you set up your iPhone, Siri may be configured. If not, you may still configure it later using the Settings app. From settings, you can modify many other things, including the Siri Voice, how it responds to notifications and incoming calls, your personal information, how Siri speaks, and much more.

It's crucial to enable the "Hey Siri" function so you may activate Siri even while your iPhone is locked in order to offer a totally hands-free experience.

Start by going to Settings from your device's home screen or app store.

Next, choose 'Siri & Search' from the list to proceed by tapping.

To enable Siri to activate using your voice, touch the toggle button located on the "Hey Siri" tile. Additionally, turn on the switch for "Press Side/Home Button for Siri" to invoke it through a button. To set up Siri, you'll need to speak a few commands; after that, just follow the instructions on your screen. Now, you may either press and hold the side or home button, or just speak to activate Siri.

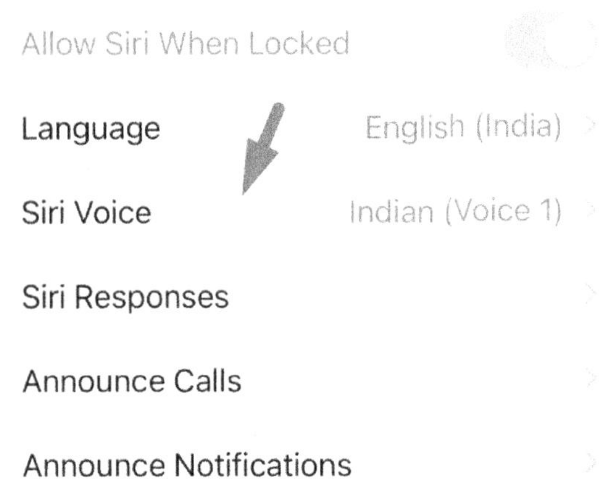

IPHONE 14 USER GUIDE

Allow Siri When Locked

Language — English (India)

Siri Voice — Indian (Voice 1)

Siri Responses

Announce Calls

Announce Notifications

> On the Siri settings page, choose the "Siri Voice" option to modify Siri's voice.

> On the Siri settings page, choose the "Siri Voice" option to modify Siri's voice.

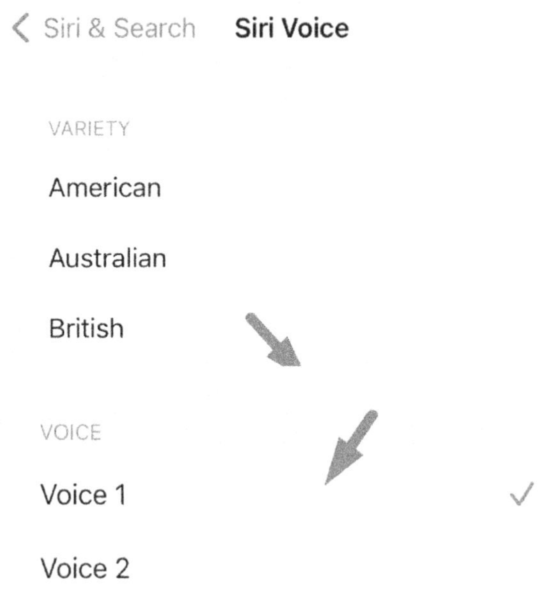

< Siri & Search   **Siri Voice**

VARIETY

American

Australian

British

VOICE

Voice 1   ✓

Voice 2

Siri Responses

Announce Calls

Announce Notifications

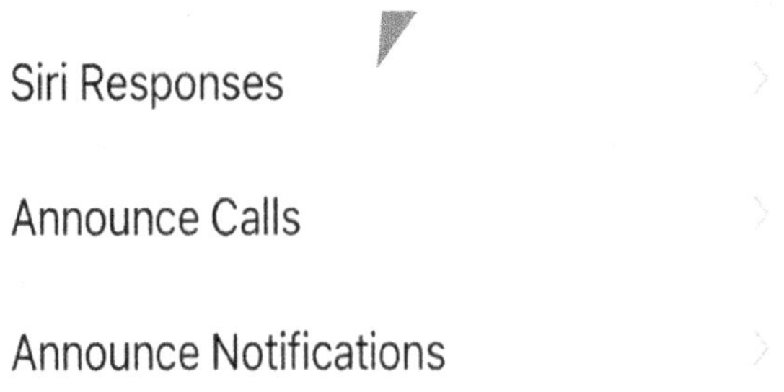

> Navigate to the 'Siri Answers' tab from the list of choices to modify Siri's responses.

-58-

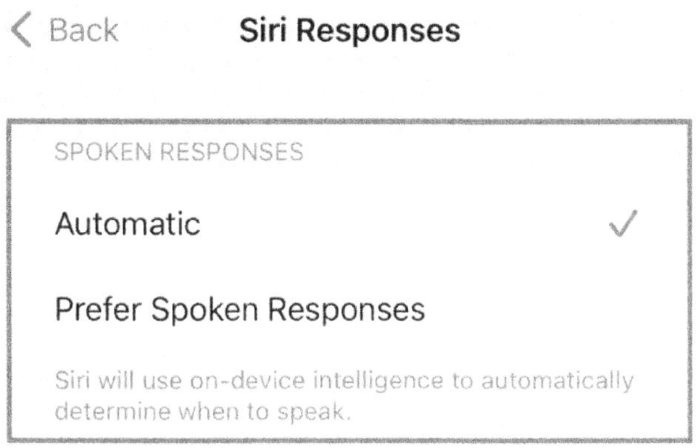

< Back   **Siri Responses**

SPOKEN RESPONSES

Automatic   ✓

Prefer Spoken Responses

Siri will use on-device intelligence to automatically determine when to speak.

Next, choose the "Prefer Spoken Responses" option if you want Siri to always respond when you dictate a command.

Next, turn on the switch next to the "Always Show Siri Subtitles" option if you want to view captions for what Siri says. Similar to the 'Always Show Speech' option, flip the switch to the 'On' position if you always want to view a transcript of your speech.

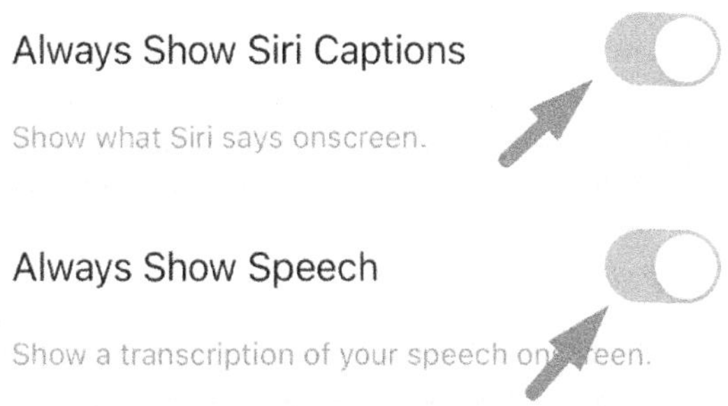

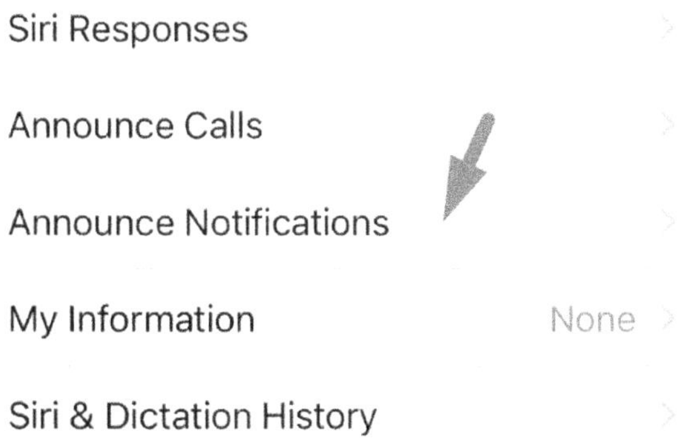

Always Show Siri Captions
Show what Siri says onscreen.

Always Show Speech
Show a transcription of your speech onscreen.

Siri Responses

Announce Calls

Announce Notifications

My Information   None

Siri & Dictation History

Tap the "Announce notifications" option to start if you want Siri to read out alerts to you.

IPHONE 14 USER GUIDE

< Back   **Announce Notifications**

Announce Notifications

Have Siri read out notifications. Siri will avoid interrupting you and will listen after reading notifications so you can respond or take actions without saying "Hey Siri". Siri will announce notifications from new apps that send Time Sensitive notifications or direct messages.

Next, choose "Announce Notifications" and then hit the toggle button.

By turning notifications "On" or "Off" from inside each app's specific settings on the same page, you can also choose which app Siri should read out notifications from.

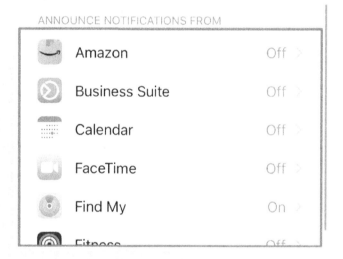

Once Siri has been set up to your preferences, you can just give orders, and it will obey.

ASK SIRI TO DO THINGS FOR YOU

As was previously noted, all you have to do is ask Siri to do a task, and it will do it correctly. Your only work will be to keep in mind the list of what it can achieve for you.

Saying "Hey Siri, Send this to Contact Name>" will cause Siri to load the information immediately in a message if you want to share anything with Siri. Instead of starting your inquiries with "Hey Siri," you may instead activate Siri by pressing the side or home button.

Siri will inform you that it can only share screenshots from the requested material and load the screenshot in the message if the requested content is not directly shareable.

Siri will ask you after the material has loaded to be sure you want to transmit it. By responding "Yes" or "No," confirm it. If you'd like, you may put a message in the available text box.

IPHONE 14 USER GUIDE

Say "Hey Siri, Call Contact Name>" to ask Siri to call a specific contact.

It will confirm by presenting you with choices if you have more than one contact name by that name or if you neglected to give the full contact name. You may select to use the last name or the position (e.g., "the first one," "second name," etc). (if available). Then, to start a call, say "Call" or hit the "Call" button.

Say "Hey Siri, set an alarm for 7 AM tomorrow/remind me to examine papers at 3 PM today/set a timer for 8 minutes" and Siri will do the desired action.

You may also use Siri to end a call by saying, "Hey Siri, end the call." Keep a note, however, since the other caller will be able to hear it.

Additionally, if you ever need a fast reminder of what Siri can do for you, just ask her, "Hey Siri, what you can do for me," and she will open a link where you can read about all of her skills.

That's all for now, guys. Using the advice above, you may quickly ask Siri to carry out a few chores while your hands are busy with other activities.

CHAPTER SIX
HOW TO CLOSE OR CLEAR APPS ON IPHONE 14

Now that you're using the iPhone 14 (or any other model in the range), you're starting to get the hang of it. Until recently, you have simply swiped up to end an app on your iPhone and carried on with your day. But it just essentially ends the app, In the backdrop, it remains open.

On your iOS device, you will ultimately need to close or clear the open applications. The first benefit is that it keeps everything moving quickly and smoothly. Additionally, the only option you have if an app suddenly stops working is to forcibly close it. It immediately ends all of the app's processes, and when you launch the app again, it begins again from scratch.

Essentially, one of the fundamentals you should understand if you are getting used to the iPhone 14 is how to forcibly close or clean the applications. But if you're unfamiliar with iOS, you can find it difficult to understand anything. If you just updated from an iPhone with a home button, you could also be experiencing issues with it. It might take some getting acclimated to all the fundamental tasks that the home button made possible.

Fortunately, it's a rather straightforward procedure, and removing an app just requires one swipe action.

Swipe up from the bottom and stop in the middle to see the app switcher and end an app.

The App Switcher will display the thumbnails for all the open applications on your smartphone. Navigate to the app preview you want to dismiss by swiping left or right.

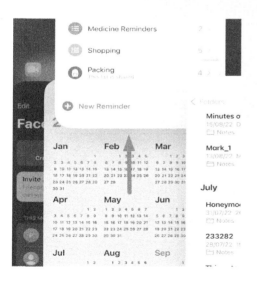

Swipe up on the app preview you want to dismiss and clear after that.

And closing the app just requires that. Unfortunately, there isn't a straightforward option to dismiss all of the open applications on your smartphone at once. To close any app, you would need to swipe it up. But you can shut many applications at once by using more than one finger. To shut three or four applications at once, you may use three fingers, which we found to be the simplest, or even four fingers, which is slower since it takes some getting used to.

The process of clearing or shutting the applications is extremely simple yet crucial. It allows you to keep RAM clean and makes your device perform quicker and more smoothly.

CHAPTER SEVEN
HOW TO FORCE RESTART IPHONE 14

The iPhone 14 line-up is now available. And now individuals are beginning to get it. The iPhone 14 is now available in all of its variations, including the 14, 14 Plus, 14 Pro, and 14 Pro Max.

If you were one of the early adopters who got their hands on the new model, you should be knowledgeable on how to force restart your phone. All ongoing processes are terminated and restarted when a process is forced to stop. You may get rid of bugs and problems you might be experiencing by forcing your iPhone to restart.

When your device is stuck in a process, won't switch off, or the touch screen isn't working properly so you can't shut it down normally, you may need to force restart it. A force restart will assist you in solving the problem, even if your device is not responding.

Finding out how to force restart the iPhone on your own might be challenging, whether you're new to the Apple ecosystem or you're switching from an iPhone with a home button. Not to fear; we've described the procedures for you and it's a straightforward process.

Note: Instead of being exclusive to iPhone 14, these instructions apply to all of the iPhone 14 models.

IPHONE 14 USER GUIDE

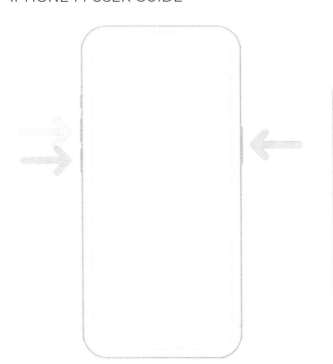

Press and release the "Volume Up" button fast to restart your iPhone 14 with a force. Press and then immediately let go of the "Volume down" button. Then, hold down the "Lock/Side" button for a moment. Keep holding the side button even if the screen with "Slide to Power Off" displays. Release the side button once you see the Apple logo on your screen. Your iPhone 14 will restart forcibly.

That's all, everyone. To restart your iPhone 14, 14, Plus, 14, Pro, or 14 Pro Max, just follow these simple steps. Many low-level faults on your iPhone may be fixed with a simple one-step technique known as "force restarting" your device.

CHAPTER EIGHT
HOW TO ENABLE AND ACTIVATE IMESSAGE ON IPHONE 14

Use the settings app to quickly activate iMessage so you can start texting with other iPhone, iPad, or Mac users.

iPhones have always been the only devices that support iMessage. Via cellular data or Wi-Fi, you may instantly exchange messages, voice memos, and media with your contacts using the iMessage service. It's one of the first things consumers want to use when setting up a new iPhone, whether they're switching from another iPhone or Android.

Although the function is automatically activated when your device boots up for the first time, there may be times when you need to do it manually. Thankfully, the procedure is much the same on iPhone 14 as it was in all earlier iterations.

The only option to resolve the iOS 16 iMessage and FaceTime activation fault that affects the recently released iPhone 14 range is to update to iOS 16.0.1. The update is only available for the iPhone 14, 14 Plus, 14 Pro, or 14 Pro Max, according to Apple. The option to upgrade your smartphone will be available to you even before the setup of your iPhone is complete. If not, go to Settings > General > Software Update to update the software.

TURN ON AND ACTIVATE THE IMESSAGE FROM THE SETTINGS APP

You may activate iMessage via the Settings app if it was either not enabled in some way during the setup of your iPhone 14 or you have previously turned it off. The procedure of enabling and activating iMessage via the Settings app is fairly simple and won't involve any work from your end or technological know-how.

IPHONE 14 USER GUIDE

Start by going to the Settings app from your device's home screen or app store.

Next, To proceed, choose 'Messages' from the list's options.

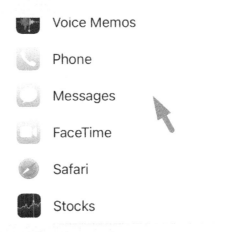

 Siri & Search

 Notifications
Banners, Sounds, Badges

iMessage

— iMessage uses wireless data to send messages between Apple Watch, iPhone, iPad and Mac. About iMessage and FaceTime & Privacy

Toggle the switch next to the "iMessage" option to the "On" position by tapping it.

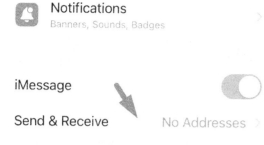

After turning it on, choose 'Send & Receive' to proceed.

Now, To receive iMessages on your smartphone, touch the cell number or email address that is provided on the screen. Your contacts will see the same address displayed.

If you have numerous addresses, choose the one you want to use when starting new chats. You are able to read and respond to iMessages sent to any address.

That's all, folks. Activating iMessage is very simple. You may now start utilizing its capabilities and effects to contact your loved ones.

CHAPTER NINE
TAKING A SCREENSHOT ON AN IPHONE 14

When you want to show off a setting or brag to your friends about a new feat in your favorite game, screenshots come in incredibly useful. On the iPhone 14, screenshotting is straightforward and uncomplicated, as it should be.

In addition, there are several approaches. You may capture a screenshot using the physical buttons, a gesture, or "Assistive Touch."

TAKE A SCREENSHOT USING PHYSICAL BUTTONS

Using the physical buttons on your iPhone is the standard and customary technique to capture a screenshot.

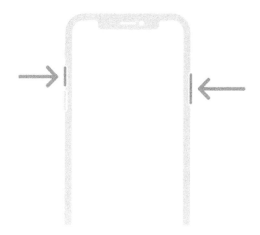

To achieve this, just press the "Lock/Side" and "Volume Up" buttons simultaneously.

After clicking, a thumbnail of the screenshot will appear on the screen. To modify, crop, or annotate it, tap on it to make it larger. Push it to the left edge of the device to dismiss it. The screenshot will be stored in Photos automatically.

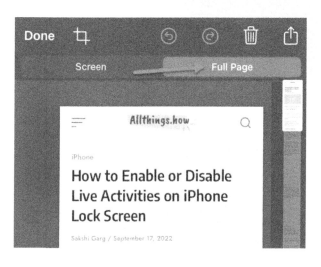

On your iPhone, you can even capture screenshots of whole web pages. Tap the thumbnail after capturing the screenshot of the website. After that, choose the "Full Page" option.

iPHONE 14 USER GUIDE

Then, you can either share it immediately using an app like Messages, Mail, etc. or save it as a PDF in files.

USE ASSISTIVE TOUCH TO TAKE SCREENSHOT

If you are unable to use both hands or push the physical buttons on your iPhone, you may easily rapidly take a screenshot using the 'Assistive Touch' function

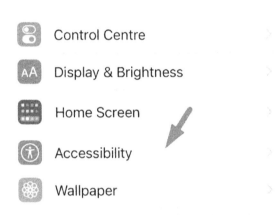

If you don't already have Assistive Touch activated, open the Settings app and choose the "Accessibility" option.

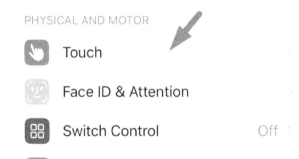

then choose "Touch" from the menu.

Finally, choose "AssistiveTouch" from the menu.

After that, toggle "AssistiveTouch" on.

Now, press the "Assistive Touch" symbol to bring it to the forefront in order to snap a screenshot in this manner.

IPHONE 14 USER GUIDE

To continue, press the "Device" option after that.

After that, choose "More" to proceed.

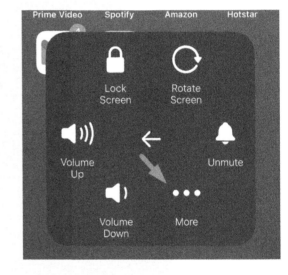

To take a screenshot of the screen that is currently visible, lastly tap the "Screenshot" option.

CHAPTER TEN
APPLE'S DYNAMIC ISLAND

The iPhone's front camera and FaceID sensor are surrounded by an interactive notch called The Dynamic Island.

It can alter look and serves as a hub for alerts, notifications, and app shortcuts. The iPhone 14 Pro versions are the only ones to support Dynamic Island.

The Dynamic Island is one of the new and noteworthy features that Apple unveiled with the iPhone 14 Pro.

It wouldn't be correct to refer to it as a rebuilt notch, but it is a fresh and inventive use of the area above the iPhone screen. The iPhone 14 Pro and iPhone 14 Pro Max are the only devices that have the feature.

The word "dynamic" describes how it is interactive. The Dynamic Island can show basic information and notifications, but it can also appear to expand in size to reveal additional information, unlike the notch, which is static and occupies space. Users have a new way to interact with their iPhones by tapping on it.

The complete information on Apple's Dynamic Island is provided here.

WHAT IS APPLE'S DYNAMIC ISLAND?

The controversial iPhone notch has been transformed into a useful new feature by Apple thanks to a complex display technology called Dynamic Island. Here is all the information you require on the new component.

The function, which is only available on the iPhone 14 Pro and uses the A16 Bionic chip, is rather obvious. is among the more significant updates to Apple's most recent flagship smartphone.

It is an island at the top of the iPhone display that interacts highly dynamically with other iPhone features, including alerts, notifications, and activities.

WHAT DOES DYNAMIC ISLAND DO?

When AirPods are attached or the device is charging, for example, the functionality is intended to surface that information. When Apple Pay is being used or the do-not-disturb mode is activated, a tiny thumbnail of the album art from Apple Music will appear on the screen.

Due to the way the pixels match the black sensors, these icons appear as an expansion on either side of the sensors, and the animations look incredibly smooth.

Dynamic Island can, however, enlarge even more to fit the larger iPhone 14 Pro display. For instance, a banner that extends from the sensors shows incoming calls and a ticking stopwatch.

If you're using Apple Pay or unlocking an app Instead of appearing in the center of the display, the Face ID indication additionally descends from the Dynamic Island.

Additionally interactive, Dynamic Island may be expanded by tapping on it. For instance, if you want to stop a voice recording or check the battery on your AirPods. A simple touch and hold may enhance ongoing activities such as Maps directions or sports scores.

When you're finished, the activity will reappear as an icon on the Dynamic Island if you swipe up to dismiss, for example, the Apple Music album art.

You will be able to connect with, to give just one example, your next Lyft trip thanks to Apple's invitation to third-party applications to participate as well. These complement the Live Activities API, which is set to launch later in 2022.

WHICH APPS SUPPORT DYNAMIC ISLAND?

Apple showcased a few during the event, but it won't be the exclusive list. Here's what we've seen so far:
- Phone
- Music
- Maps
- Wallet

- Voice Memos
- Clock
- Lyft

Other capabilities include connection for headphones (perhaps only AirPods), do not disturb mode, quiet mode, charge status, and more. Along with sporting results for your preferred teams, team badges on each side of the sensors will display the results.

HOW DOES DYNAMIC ISLAND WORK?

All of this is made possible by the hardware modifications made to the iPhone 14 Pro, and it consists of two parts: The design of the display itself is important first. The proximity sensors have relocated behind the display for the first time ever, and the orientation of the cutout has been altered to appear neater. Second, since the A16 Bionic processor is there, it is conceivable (and only on iPhone 14 Pro versions).

IT SHOWS WHEN YOUR IPHONE 14 PRO IS LOCKED OR UNLOCKED

The Dynamic Island alerts you when your phone unlocks by gently expanding and displaying a straightforward lock/unlock animation.

IT DISPLAYS THE CHARGING STATUS OF YOUR IPHONE

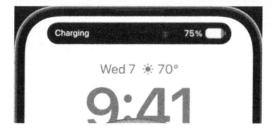

The Dynamic Island stretches to provide a subtle signal that you are charging your phone and displays your battery level when you put your iPhone in to charge.

IPHONE 14 USER GUIDE

YOU CAN CONTROL PHONE CALLS

When you get a call on the Dynamic Island, you'll receive the usual notice that gives you the option to pick up or hang up. When you're through with a call, you can easily end it by tapping the Dynamic Island and selecting the "Hang Up" option without leaving the app you're now using. In the past, in order to hang up a call, you had to hit the phone call alert in the upper left corner of the screen.

IT SHOWS YOU WHEN YOU CONNECT TO AIRPODS

Instead of standalone notifications that showed in various locations around iOS, the paired AirPods notice is going to the Dynamic Island, which has evolved into a focal point for alerts and notifications. You can see which pair of AirPods are paired as well as the battery level by tapping on the Dynamic Island.

IT CAN DISPLAY MUSIC YOU'RE LISTENING TO AND MEDIA CONTROLS

While playing music, album art appears on the Dynamic Island. You may touch on the Dynamic Island to access music controls like play/pause/previous/next. It also offers the option of turning on AirPlay.

For basic audio controls, you won't need to restart your music streaming application or slide down from the top-right corner of the screen to see the control center. Additionally, if you currently use a widget to manage your music, the Dynamic Island allows you to regain that area.

YOU CAN GET TRAVEL ALERTS

-80-

Instead of a notice that directs you to the app itself, useful, dynamic information and updates from applications may be shown on the Dynamic Island. The flight number, whether the trip is on schedule, and how much luggage the passenger has checked in are all shown when you tap on the Dynamic Island for travel information.

IT CAN GIVE YOU TURN-BY-TURN DIRECTIONS

You'll get a notification for an impending driving direction step from the Dynamic Island while the Maps app is running in the background, further centralizing where you'll discover essential information. When it's time to go on, the alarm will enlarge, which draws your attention.

IT SHOWS YOU AIRDROP NOTIFICATIONS AND STATUS

The Dynamic Island's new AirDrop notification lets you track the download or upload's progress. To check the status of the AirDrop in further detail, touch the Dynamic Island.

IT SHOWS THE STATUS OF RIDE-SHARE SERVICES LIKE LYFT

The Dynamic Island will repeatedly display how long it will take your ride-sharing vehicle to arrive. The name and picture of the driver, the driver's rating, the brand of the automobile, and license plate details are all available by tapping on the island.

IT SHOWS WHEN FACEID IS BEING USED

In order to further assemble alerts and notifications at the top of the iPhone, the FaceID animation stretches from the Dynamic Island.

It can show you multiple shortcuts for things you're doing on your iPhone

The Dynamic Island can display what processes are active in the background, and you can touch on the corresponding icons to quickly access the corresponding program.

The Island, for instance, will display a timer symbol if music is playing and a timer is set. You may access these applications by clicking on each icon.

WHICH PHONES ARE COMPATIBLE WITH DYNAMIC ISLAND?

The unique location of the front-facing camera and Face ID sensors, which are now embedded inside the display rather than as a chunk cut out from the top of the device, makes it feasible for the new Dynamic Island, which is exclusive to the iPhone 14 Pro and iPhone 14 Pro Max.

CHAPTER ELEVEN
THE IPHONE 14 CAN CONNECT TO SATELLITES

The iPhone 14 and iPhone 14 Pro utilise several clever new safety measures, such as the capability to contact with friends and emergency services through satellite communication. This is what we do know.

HOW TO SOLVE THE SATELLITE ISSUE

Unlike cell towers, which stay in the same place, satellites move quickly around the sky (relatively speaking). There is not much bandwidth available for portable satellite communication. Although satellite internet is available, it requires large antennas and a lot of electricity.

This problem is undoubtedly familiar to you if you've ever used a satellite phone to make or receive a call. Due to the limited bandwidth, dropouts are frequent and audio quality is often poor. Communication must be severely compressed, sometimes to the point of being unhearable.

With the "Emergency SOS via Satellite" function in the iPhone 14 and iPhone 14 Pro, Apple attempted to address both of these issues. You must have a good view of the sky outside in order to utilize the service, according to the provider. Your iPhone will indicate which way you should position it to establish and maintain a connection to make the procedure simpler.

Emergency SOS via Satellite can only send text messages in order to get around bandwidth restrictions. To reduce text messages three times as much as uncompressed data, Apple created a text compression method. Under ideal circumstances, a single text message can be transmitted in less than 15 seconds, but under "light foliage," it may take more than a minute.

A text-only strategy presented Apple with even another challenge. Some emergency receivers only take voice calls; others only accept text messages. It sounds like a manned version of Emergency SOS through Siri on the Apple Watch, but Apple created infrastructure to send messages via "emergency relay centers with Apple-trained professionals who call for aid on your behalf."

FOR EMERGENCY SERVICES AND FRIENDS

Your iPhone will ask you a few questions to determine your circumstances before you connect to an emergency operator. Your responses to those queries, together with your location, your medical ID, and the battery level at the time you made the connection with an emergency responder.

The goal is to provide as much information as is required to aid emergency responders in finding you and locating you. Keep your iPhone aimed towards a satellite to retain a connection; if necessary, a notice should show up on the screen to let you pick up where you left off with the operators.

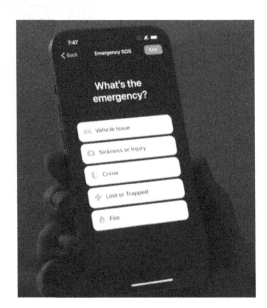

Although the power requirements of satellite communication are not yet known, if GPS performance is any indication, it will certainly deplete your battery far more quickly than making a call via the cellular network. Your iPhone utilizes GPS to get a satellite-based position fix, and this feature has historically been a huge battery drainer.

Apple's satellite technology may be used in conjunction with the Find My service to inform friends and family of your whereabouts as well as for direct communication with emergency services. Apple's Find My geolocation software is designed for locating people, devices, and items with AirTags.

At this early point, Apple has been reticent to discuss this aspect of the service, however it is assumed that it will only function with other Find My users (requiring an Apple ID to function). The service seems to need manual updating rather than operating automatically in an early screenshot, but we won't know for sure until the complete release.

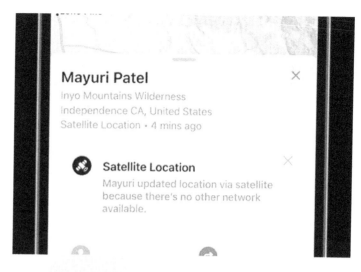

HOW MUCH WILL SOS VIA SATELLITE COST?

The persistent concern about recurring costs may be the most mystifying aspect of the iPhone 14 and iPhone 14 Pro satellite capability. During the iPhone 14 launch, Apple said that the service will be "free for two years" with compatible devices, but it didn't elaborate on how much consumers would have to pay after that.

For voice plans with unlimited talk and text and worldwide coverage, prices typically start at roughly $50 per month.

Apple's service should be less expensive than others since it just supports text. Though it's evident that there's more involved than just infrastructure expenditures in this case given that Apple has indicated that "Apple trained professionals" would be employed to make the service work.

In the same way that it has done with its Private Relay and Hide My Email services, Apple may include the service into iCloud+. Given the need to hire and educate professionals, it seems implausible. It would seem more logical to anticipate that the service will have a monthly or annual price, however it is unclear how much that would cost.

Satellite phone services provide worldwide coverage, however Apple is not currently providing such services. Before the two years are up, Apple could want to concentrate on expanding the service to additional locations so that it will be a more alluring offering when it comes time to renew.

For people who don't always need satellite communication, having a subscription that is pre-paid for the month could work. Instead of paying for satellite SOS (and Find My Location) services throughout the year, you might pay for the months you're most likely to need them by spending a few months hiking in the summer and a few months skiing in the winter.

CHAPTER TWELVE
HOW CRASH DETECTION WORKS AND HOW TO TURN IT OFF

Crash Detection is a brand-new security feature that Apple unveiled with the release of the iPhone 14 and new Apple Watch models. To find out what it is, how it works, and how to switch it on and off, keep reading.

CRASH DETECTION: WHAT IS IT?

The goal of collision detection is to identify serious auto accidents in passenger vehicles. Your Apple device may automatically contact emergency services on your behalf if you have a serious collision and become unresponsive.

HOW DOES CRASH DETECTION WORK?

Your iPhone or Apple Watch uses a motion sensor with a high dynamic range gyroscope and high-g accelerometer, GPS, barometer, microphone, and powerful motion algorithms to jointly identify crashes when you're operating a motor vehicle or riding as a passenger.

A 10-second alarm appears on the screen of your Apple Watch or iPhone when a major accident is discovered. If you're still awake, you may swipe the screen to phone 911 right away or turn off the alarm if you don't need to get in touch with them. A 10-second countdown will begin if you don't touch your Apple iPhone after 10 seconds. Emergency services are notified after it is over.

IPHONE 14 USER GUIDE

The following Siri voice message starts playing on repeat five seconds after the emergency services answer: "The owner of this iPhone was in a catastrophic vehicle collision and is not answering to their phone." Following that, Siri will communicate your approximate location—complete with latitude, longitude, and a search radius—to emergency services as well as via the device's speaker.

The first time the message is broadcast, it is the loudest; subsequent repeats are at a lower intensity in case you can communicate with emergency personnel. Until you hit the Stop Recorded Message button or the call is disconnected, the message keeps repeating.

A slider will display on the Lock Screen for easy access to your Medical ID information, and if you have established emergency contacts, they will also be alerted of the accident after a further 10-second countdown.

WHAT KINDS OF CRASHES CAN IT DETECT?

Apple claims that Crash Detection can identify rollovers, rear-end accidents, side hits, and front impacts.

WHICH APPLE DEVICES SUPPORT CRASH DETECTION?

The following Apple Watch and iPhone devices may use crash detection.

- iPhone 14
- iPhone 14 Plus
- iPhone 14 Pro

- iPhone 14 Pro Max
- Apple Watch SE (2nd Generation)
- Apple Watch Series 8
- Apple Watch Ultra

HOW TO ENABLE CRASH DETECTION

There is no setup needed. On the supported devices listed above, crash detection is turned on by default, so there is nothing you need to do. You may turn off the function by doing the things listed below if you're worried that it could inadvertently record an accident and contact emergency personnel.

Activating Crash Detection

- Start your iPhone's Settings application.
- Tap through to Emergency SOS once you scroll down.
- Turn off the switch next to Call After Severe Crash under "Crash Detection."
- Simply turn the switch back on in Settings if you ever wish to activate the function.

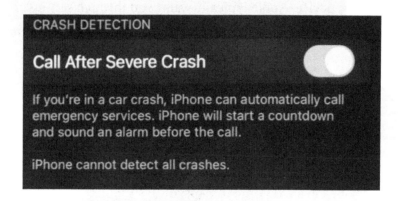

CHAPTER THIRTEEN
HOW TO USE APPLE PAY

The simplest method to begin using contactless payments on an iPhone is using Apple Pay. The digital wallet service is not only free to use, but it is also pre-installed on your phone.

To use Apple Pay, you must connect a digital debit or credit card to the Wallet app. Once you're set up, you can use Apple Pay to make purchases offline, online, and in apps without needing to carry any cash or additional wallets or actual credit cards. Through the Messages app, you may now give money to loved ones using Apple Pay. In certain places, you can even use Apple Pay to pay for public transportation in replacement of a transit pass.

When you first turn on a new device, Apple typically walks you through setting up Apple Pay. However, if for some reason you didn't, here is all the information you want to get going as well as some suggestions on how to use it.

SETTING UP APPLE PAY

Checking to see whether your device is compatible is always the first step when setting up any gadget. With the exception of the iPhone 5S, it refers to any iPhone model that has Touch ID or Face ID. Additionally, all Apple Watch models since the Series 1 and certain iPad models with Touch ID or Face ID support Apple Pay. Apple Pay could also be supported on Mac models with Touch ID. In any case, you may check here to see whether your particular gadget qualifies.

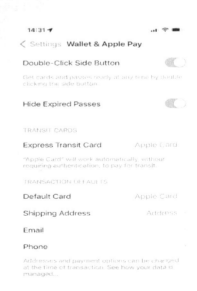

A compatible credit or debit card and an Apple ID that is signed into iCloud are the next items you'll need. During the setup process, you may be prompted to sign in if you aren't already.

ON AN IPHONE

- Open the Wallet app on your iPhone to add a card.
- Hit the "+" key.
- Choose a Debit or Credit Card from the list of available cards.
- Select Continue.
- From this point, you can either take a photo of your actual card or manually input the information.
- Check your card's information.
- Then click I accept the terms and conditions.

You may decide whether the newly added card should be your default card.

You will now be requested to add your card to your Apple Watch, if you have one. You may finish this later.

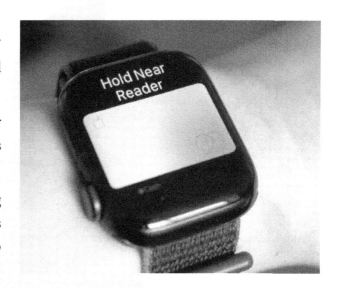

Once you've completed this on an iPhone, adding the identity cards to other Apple devices is significantly simpler. We advise adding your cards to the iPhone first for this reason

ON AN APPLE WATCH

Make sure a passcode is enabled before using Apple Pay on your Apple Watch. You must comply with these criteria in order to utilize Apple Pay on the Watch.

On the Apple Watch, add a card as follows:

Get your iPhone's Watch app open.

Click on the My Watch tab.

To access Wallet & Apple Pay, scroll down.

You may proceed in the same manner as before from here.

You should find them displayed under Other cards on your iPhone if you've added any cards earlier. To add a card to your Watch, tap the Add button next to it.

After checking your card information, you're done!

IPHONE 14 USER GUIDE

ON AN IPAD OR MAC

You may transfer money using the Messages app on the iPad and Mac using Apple Pay to make purchases online and in applications.

To configure Apple Pay on a Mac or iPad:

- Open System Preferences (Mac) or Settings (iPad) (Mac).
- Choose Apple Pay & Wallet.
- Choose Add Card.
- Any Previous Cards will therefore be shown under Available Cards from this point forward.
- Choose the cards you wish to include.
- Choose Continue.
- After checking your card information, you're done!

USING APPLE PAY

Take some time to adjust your settings before you start pulling out your phone to pay. Make sure to choose which card is your default if you have several cards stored in the Wallet app. Additionally, you must activate the Apple Cash card in order to give and receive money via the Messages app.

Apple Pay needs two-factor authentication by default, which is why you need a smartphone with Touch ID or Face ID compatibility or to activate a passcode on your Watch. However, you may alternatively designate an Express Transit Card if you'd rather get around that.

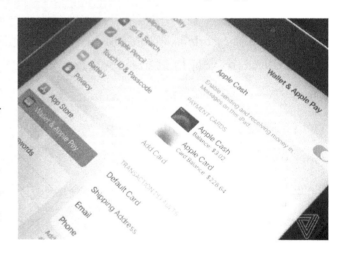

On the iPhone, Apple Watch, and iPad, go to Settings > Wallet & Apple Pay to get to these settings. Go to System Preferences > Wallet & Apple Pay on the Mac.

There are several methods to use Apple Pay, but the majority of users will probably do it on their iPhones or Apple Watches. You may use Apple Pay everywhere you see these symbols, so long as you see them.

PAY USING AN IPHONE OR APPLE WATCH

Double-click the side button on an iPhone with Face ID to make a payment. The phone will then try to verify you using Face ID as the default card will then display. You may be prompted to enter your passcode if it is unable to. With Touch ID phones, the procedure is the same, but you authenticate by placing your finger on the Touch ID button.

You may also double-click the side button on the Apple Watch to display your chosen card. You may then hold your phone or Watch over the contactless payment scanner when that is finished.

All you need to do to complete a purchase in an app or on a website in Safari is click the Apple Pay button. Depending on the device, you will then be prompted to confirm your identity using Touch ID or Face ID. Once you've done that, double check your payment details and the card you wish to use.

SEND MONEY WITH MESSAGES

You must be in the US and have the Apple Cash card activated in order to transfer money via the Messages app, as must the recipient. On your iPhone, go to Settings > Wallet & Apple Pay > Apple Cash to do this action. You may then connect your bank account to fund the card after that is complete. You may also transfer any remaining money on the card back to your bank.

You just need to press the Apple Pay or Apple Cash symbol in a specific iMessage discussion once Apple Cash is configured. Then you may request or make money after being requested to enter an amount.

That's all there really is to it. Happy shopping.

HOW TO OPEN APPLE PAY FROM IPHONE LOCK SCREEN IN 3 EASY STEPS (IOS 16)

IPHONE 14 USER GUIDE

Access your credit cards from the Lock Screen with only a double-click by using an Apple Wallet shortcut.

Did you know that the iPhone's lock screen can be used to access Apple Wallet? You can instantly access Apple Pay as well as the passes and loyalty cards stored in your Apple Wallet when this Wallet app feature is activated, allowing you to be ready to make a purchase or pull up your boarding pass while you're in line.

WHY YOU'LL LOVE THIS TIP

Get easy access to your no-contact payment options.

Quickly access your concert tickets, discount cards, boarding passes, and more.

ACCESS APPLE WALLET FROM THE LOCK SCREEN

Make sure you've gone through how to set up Apple Pay and how to update your Apple Pay Payment options before we learn how to set up and then access Apple Wallet from the Lock screen.

Open the Settings app.

Wallet & Apple Pay may be accessed by scrolling down and tapping it.

You will find a toggle named Double-Click Side Button on the majority of contemporary iPhone models. Ensure that this is turned on (green).

If you use an iPhone with a Home button, you could instead see Double-Click Home Button.

Now, all you have to do to open the Wallet app on an iPhone's Lock Screen is double-click either the side button or the Home button. This will bring up a screen similar to this one:

Holding your iPhone close to the card reader will enable you to complete the transaction. It should be noted that this won't launch the app if you don't already have a credit or debit card set up for Apple Pay.

Your days of searching through your wallet or handbag for your credit cards are past, and you should always be able to use Apple Pay from your lock screen.

CHAPTER FOURTEEN
HOW TO MASTER THE IPHONE 14 PRO & IPHONE 14 PRO MAX CAMERA

The brand-new iPhone 14 Pro and iPhone 14 Pro Max both include a Camera and app that are highly powerful. Here are some tips for taking the greatest pictures.

Three cameras are available on the back of the iPhone 14 Pro and iPhone 14 Pro Max. In addition to a 12MP ultra-wide lens and a 12MP telephoto lens, the device has a 48MP wide-angle main camera.

Apple improved the camera this year with a revamped Photonic Engine, Action Mode, and other features. Let's go through the ever-expanding list of features available on Apple's most recent pro iPhones.

VOLUME BUTTONS

Starting with the Camera app's controls, touching the screen isn't always the best option. The phone may move with only one touch, accidentally distorting your picture in the process.

This problem is addressed by Apple by enabling the volume buttons to also serve as convenient shutter controls.

The volume up or down button will automatically take a picture when pressed. Holding either button initiates video capture. The video capture will end once the button has been released.

You may modify this behavior by going into Settings. You may enable burst capture by going to Settings > Camera.

IPHONE 14 USER GUIDE

When activated, holding the volume up button will begin taking bursts of images instead. When you release the button, the images will no longer be taken.

CONTROLLING THE ZOOM

There are four optical-quality zoom settings available on the iPhone 14 Pro and iPhone 14 Pro Max. 5x exist, using the ultra-wide lens, one time, using the main wide lens, twice, using the main lens, and three times, using the telephoto lens.

The theory is that Apple can accomplish a 2X magnification without noticeably sacrificing quality by using the center 12MP from the bigger 48MP sensor. If we were to go very scientific, there is some quality loss since Apple is unable to apply pixel binning when this occurs, which means the smaller pixels collect a little less light.

Nevertheless, it will result in better pictures than using digital zoom. Instead of changing the optics, digital zoom works by cropping the image, which results in a lower-resolution image.

By touching the circular indications above the shutter button, you may choose between these pre-set zoom settings. However, there is another little-known method that allows you to hold and glide your finger up or down when you touch any of them.

With this motion, a zoom wheel will appear so you may choose the degree of magnification you want, up to 15X digital zoom. The corresponding focal length is shown below the preset zoom settings.

Once finished, you may either wait a little while for it to vanish on its own, or swipe the wheel in the direction of the shutter to get rid of it. Pinching in or out is the other choice, however, it is less practical and blocks the screen.

Bring your phone extremely near to your subject to activate Macro Mode if that's what you want to capture. It is able to focus millimeters away from the lens. You may turn on a toggle in Settings to prevent this from happening automatically.

APPLE QUICKTAKE

QuickTake makes it simple to immediately snap images, movies, or burst shots. If you tap the white shutter button, it will take a picture for you; if you hold it, it will take a video. avoiding the need to switch to video mode.

As long as you hold down the shutter button, the video will continue to be recorded; once you let go, it will cease. Slide the shutter button to the right to lock it in video mode and avoid holding the button down if you want to record for a lengthy.

You may press the button and then quickly pull it to the left to take burst pictures. Now, as long as you keep holding it, it will take burst images. In the original shutter button circle, a counter will show how many pictures were taken.

BONUS CONTROLS

Additional controls that alter dependent on your shooting scenario are situated at the top of the phone, on each side of the Dynamic Island. The flash symbol, the icon for Night Mode, the RAW or ProRes indication, and the toggle for Live Photo may all be visible.

In addition to this, you may access more settings above the shutter by tapping the carrot arrow in the exact middle of the screen. The icons for the flash control (auto, off, on), Night Mode (appearing only in low light), Live Photo (on, off), Photographic Styles (choose one of five styles), aspect ratio (4:3, 1:1, 16:9), exposure correction, timer, and filters are among those present.

With the iPhone 14 Pro, Night Mode is still available and mimics the effects of a slow shutter speed. On a conventional camera, keeping the shutter open for a longer period of time will let more light reach the sensor, enabling you to shoot pictures in low light. You simply cannot move the camera while the picture is being shot, and that is the secret.

When it recognizes a scenario with little light, Night Mode will automatically activate. In accordance with how much the phone and subject move, it will alter the shutter time. The iPhone could have a two- or three-second shutter restriction for freehand shots. The shutter may remain open for up to 30 seconds when supported by a tripod.

You may manually change this shutter length by moving it left and right by clicking the Night Mode symbol. However, if the phone is moving too quickly, it won't offer you a choice for 10, 15, or 30 seconds.

Although you may add filters after you capture a picture, the iPhone 13 series introduces Photographic Styles. These may be modified to your shooting preferences and are added at the time of capture. By hitting the undo arrow on the right side, you may undo changes you've made if you don't like them.

In some circumstances, it may be best to shoot in RAW. These uncompressed photographs use a lot more storage space but provide more clarity, particularly in the highlights and shadows.

You may take 12MP RAW images with the iPhone 14 Pro or use the complete 48MP sensor. The picture may have additional grain if the whole sensor is used. Therefore, use RAW mode only if you are aware of the risks.

VIDEO RECORDING ON IPHONE 14 PRO

The iPhone 14 Pro comes with four video settings from Apple, kind of. There includes ordinary video mode, timelapse, slo-mo, and cinematic.

The resolution and frame rate will be shown in the top-left corner when you enter video mode. To alter, just tap one of the numbers. Different recording resolutions and framerates will be available based on what the phone can capture depending on your video mode.

Apple improved Cinematic Mode this year so that it can record video at up to 4K resolution on the iPhone 14 Pro. Prior to this, it could only be in HD 1080P. You may record at 24 or 30 frames per second while filming in 4K.

If you've never heard of Cinematic Mode, it is essentially Portrait Mode for videos. It will blur the background of the image while keeping your topic in focus. After you've finished filming, you may change the focus as required to accentuate the things you desire.

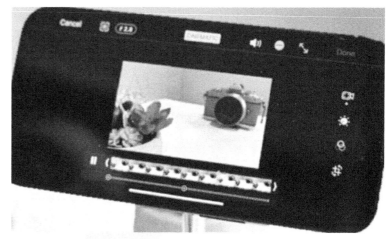

You have control over both the subject and the simulated aperture when editing. Similar to Portrait Mode, you may vary how much of the backdrop is obscured by changing the aperture.

Action Mode is new for iPhone 14 Pro owners. You may record smooth video while moving quickly in action mode. Imagine it as a highly stabilized video that you would typically shoot with just your iPhone and a gimbal.

Simply put your phone in video mode to activate this mode, and you'll see a new icon at the top. When activated, you may walk about while still capturing fluid video, and it has the appearance of a guy sprinting.

The video in Action Mode is somewhat cropped and has a maximum resolution of 2.8K. For your phone to apply post-shooting stabilization, the crop is required.

Action Mode requires a lot of light, however there is an option that makes it possible to use it in low light. The stabilizing impact won't be as strong with this on.

GET SHOOTING

Although iPhone users always have the choice to utilize third-party camera apps, many continue to stay with Apple's pre-installed options. Because of this, it is crucial that Apple keeps introducing new capabilities to the built-in Camera app.

The Camera app has you covered whether you're shooting in Portrait Mode, taking live shots, or creating artistic Cinematic Mode videos.

CHAPTER FIFTEEN
IOS 16

Beginning in September, Apple unveiled the iPhone 14, 14 Plus, 14 Pro, and 14 Pro Max, all of which ship with the most recent iOS pre-installed. Owners of current and a variety of previous iPhone models may also download iOS 16.

Here is all the information you need to know about iOS 16, including all the features and the release date.

LAUNCH OF IOS 16

At its annual developer conference, WWDC, in June, Apple revealed the next significant software upgrade for the iPhone. Betas for developers and the general public were then issued.

Then, on September 12, 2022, iOS 16 became fully accessible to the whole public.

Have you ever been a bit bored with the features on your iPhone? To feel excited, you download one or two new apps, but the emotion quickly wears off. Prepare yourself to get a brand-new phone without having to purchase one.

There's still something for you if you're not buying anything from the iPhone 14 selection that was shown on stage at the Apple event last week. iOS 16 from Apple is now available. The familiar lock screen is no longer as familiar, which instantly changes how it appears. The date and time may now be changed, and you can even add dimension to the snapshot you save there. Not to add, there are new widgets methods.

If you look into iOS 16 a little further, you'll discover that iMessages includes a ton of new capabilities, like the capacity to modify and even remember text messages. Siri's dictation abilities have advanced to the point that it now automatically adds emojis and handles punctuation. You can now link your transport passes to Maps, which is a lot better tool for traveling.

In iOS 16, there is a ton more to discover. Continue reading for instructions on downloading iOS 16 and suggestions on how to be prepared for the next updates.

CAN MY IPHONE RUN IOS 16?

- 2022 iPhone SE
- 2022 iPhone SE

Even though your outdated iPhone is still functional, it may not be able to run iOS 16. But if you own one of these iPhones, you're safe:

- 14 series
- iPhone 13 mini
- iPhone 13 Pro
- iPhone 13 Pro Max
- iPhone SE (2022)
- iPhone 12 mini
- iPhone 12
- iPhone 12 Pro
- iPhone 12 Pro Max
- iPhone SE (2020)
- iPhone 11 Pro Max
- iPhone 11 Pro
- iPhone 11
- iPhone XS Max
- iPhone XS
- iPhone XR
- iPhone X
- iPhone 8 Plus
- iPhone 8

IOS 16: FEATURES

Here are some of the features that iOS 16 adds to your iPhone.

1) ADD WIDGETS TO MULTIPLE LOCK SCREENS

With the release of iOS 16, the lock screen has undergone a significant overhaul. You can now add your own widgets to it, and you may store several lock screen setups and switch between them as needed. To manage your lock screens and customize the wallpaper and various widget combinations shown on each one, press and hold on to the display when the lock screen is active. Then choose Customize.

2) CHANGE HOW LOCK SCREEN NOTIFICATIONS ARE SHOWN

You now have additional options over how alerts are shown when they come, staying with the lock screen: Select Notifications from the Settings menu, then pick one of three options. The Stack option collects alerts from the same app, but the Count option just displays the total amount of unread notifications. The last option, List, is how alerts were shown in earlier iOS iterations.

3) EDIT AND UNSEND MESSAGES

The ability to amend texts sent over iMessage through the Messages app (up to five revisions per message) or to totally unsend messages is one of the most notable new capabilities this time around (you get two minutes of thinking time). When you long press a message in a chat, the Edit and Undo Send options will show up on the screen, allowing you to make changes to the message you just sent or send it back to the ether. Receivers who don't have iOS 16 won't see your unsent or unedited messages, but they will instead get a fresh message for every modification.

4) RESTORE JUST-DELETED TEXTS

The Communications app now offers the option to restore conversations that have been erased within the past 30 days. This feature works somewhat similarly to the Mac's Trash folder and gives you an additional measure of protection before messages are permanently lost. Open the Messages app, choose Edit from the menu, and then select Show Recently Deleted. Any of the mentioned message threads may be restored from the next screen, or you can permanently delete them.

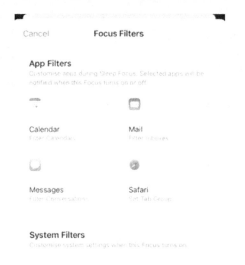

5) SET UP FILTERS IN FOCUS MODE

In iOS 16, there are several improvements to the Focus function, including the option to configure filters in specific applications like Messages, Mail, and Calendar to reduce the number of distractions you have to deal with. By selecting Focus from the Settings menu, you may choose Add Filter to achieve precisely that: For instance, you may simply restrict the number of calendars shown in the Calendar app or hide certain email accounts in the Mail app.

6) SCHEDULE THE SENDING OF EMAILS

You may now schedule emails in Mail, and it also includes an undo send option similar to Messages (active for up to 30 seconds after sending). When you press and hold on the blue arrow symbol in the upper right corner of an email after you've completed writing it, the following choices will appear: Pick one of the suggested options, or press Send Later... to set your own time (just make sure your phone is online at the time).

7) LOCK DOWN YOUR IPHONE

Apple described the lockdown mode in iOS 16 as an "optional, severe protection" approach that most users will never need: It severely reduces the area that hackers may target while also reducing parts of iOS's capabilities (incoming FaceTime calls and attachments in Messages are blocked, for example). By selecting Privacy & Security from the Settings menu, followed by Lockdown Mode, you can get to it.

8) SHARE TAB GROUPS IN SAFARI

This may not be something you use all the time, but it could come in handy once in a while: Through Safari on iOS, you can now share sets of tabs with other users if you're working on a project or doing research with them. Open a tab group and hit the share icon in the upper right corner to choose a few contacts. Note that tab groups may now have their own unique start pages as well as pinned tabs.

IPHONE 14 USER GUIDE

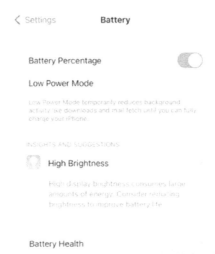

9) SEE THE BATTERY LEVEL PERCENTAGE

Once again available, the battery percentage option makes it simpler to see how much energy is still in the tank. Go to Settings, choose Battery, then turn on the Battery % toggle switch to display the current percentage level on the battery icon in the status bar. Notably, the iOS 16.1 update apparently adds the battery % option to any iPhone with Face ID, even though certain iPhones were left out when iOS 16 first debuted.

10) PLAN ROUTES WITH MULTIPLE STOPS IN APPLE MAPS

Since iOS 16 has been loaded, you may now construct multi-stop routes in Apple Maps, something Google Maps has offered for many years. As you would normally, set up some paths between two points, and you'll see the new Add Stop button. One set of instructions may include up to 15 stops, and you can reorder them by dragging and touching the handles on the right side of the screen.

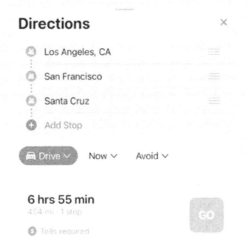

11) TRACK YOUR FITNESS WITHOUT AN APPLE WATCH

The Apple Fitness app no longer requires an Apple Watch to track your activities; if you don't have the wearable, it will utilize the motion sensors on your iPhone to measure your steps and the distances you walk. It may even be compatible with certain third-party exercise programs. Even though the capability is limited compared to an Apple Watch, it can nevertheless figure out how many calories you burn each day.

12) RUN A SAFETY CHECK

Apple has introduced a new tool called Safety Check for individuals who are "in domestic or intimate partner abuse situations." It provides a rapid overview of the users and applications linked to your account and lets you remove permissions like those allowing access to your location. We hope you never need to use this option, but if you do, you can locate it in Settings under Privacy & Security: You may examine your settings or reset them by tapping Safety Check.

13) CUT OUT OBJECTS FROM PHOTOS

One of iOS 16's most useful new features is accessible in a number of applications, including Photos and Safari. It enables you to isolate topics in photos from their surroundings: When an image is open on the screen, press and hold on to the primary subject. If iOS can recognize the boundaries of the item correctly, a faint white light will emerge around it. Then, by pressing the Copy or the Share button, you may transmit the item that was cut out to another location.

14) TRACK YOUR MEDICATIONS

With the release of iOS 16, the Apple Health app can now monitor your prescriptions. You must input all information manually, but it is a useful tool for making sure you take your meds as prescribed each day. Select Add a Medication: from the Health app's Browse menu after selecting Medications. Its name and how often you take it will be requested of you.

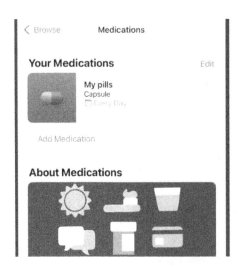

15) ADD HAPTIC FEEDBACK TO THE KEYBOARD

In iOS 16, the native iPhone keyboard receives haptic feedback for the first time, allowing you to have extremely light vibrations accompanying every press on the device. This feature is popular with users since it allows them to be absolutely certain that their inputs are being recorded. Open the Settings screen on your iPhone, choose Sounds & Haptics and Keyboard Feedback, and then toggle on the Haptic toggle button that displays to activate the option.

16) SHARE PHOTOS MORE EASILY

A new feature called iCloud Shared Photo Library will be available soon to provide an easier, more practical way of sharing photos and videos with a specific group of people, though it is still listed as "coming soon" at the time of writing (Apple claims it will be available before the end of the year) (like close family members). Anyone with whom you have shared a library may edit, remove, and watch the images and videos you provide.

HOW TO GET IOS 16 ON YOUR OLD IPHONE

You may either wait for your phone to pop up an alert telling you that iOS 16 is ready, or you can manually force the upgrade. (Backing up your data is usually a good idea before updating.)

Select Settings > General > Software Update from the menu. Either you'll be instantly sent to Download and Install or you'll need to press Upgrade to iOS 16 at the bottom of the page before you can Download and Install. In either case, you'll be asked to enter your iPhone's password if it has one. Accept Apple's conditions, then wait thereafter.

The message Update Requested appears on the screen, suggesting that Apple has added you to the download queue. Once iOS 16 begins downloading, you'll see a time estimate meter at the top; how long you have to wait depends on how many people are attempting to update. The next step is to reset your phone, which may take a few minutes.

You'll see a notification requesting that you temporarily uninstall applications if your device does not have enough space for the latest iOS. When the installation is complete, press Continue, and the applications will be restored. Hit Cancel and manually uninstall any programs you don't trust before upgrading.

Install Tonight, which will install iOS 16 as you sleep, is another alternative if you require your phone during the day while your device is charging.

HOW TO CUSTOMISE YOUR IPHONE LOCK SCREEN WALLPAPER, FONT, COLOUR, AND WIDGETS

The option to customize an iPhone lock screen was added with iOS 16, which Apple launched in September. Not just switching out your wallpaper, either. Now, you can customize a lock screen's whole appearance, including the colors, fonts, and widgets that show data from your preferred applications. If you want to vary your look, you can also make many lock screens to choose from. Additionally, if you've created an Apple Focus, you can connect it to your own personalized lock screen. The concept is that you may switch to a certain custom lock screen with a Focus activated when you wish to lessen distractions like alerts.

Confused? Not to worry. This post has provided step-by-step instructions on how to customize your iPhone lock screen.

HOW TO CREATE A CUSTOM IPHONE LOCK SCREEN

Below, we outline everything you can do to customize your lock screen, from beginning to end. We go into further depth about certain topics in this article, such as how to add a picture or widgets to your lock screen.

Make a new lock screen:

Touch and hold the Customize button until it shows on your lock screen.

At the bottom of the screen, tap the Add New or + button.

The collection of lock screen wallpapers will show up.

There is a gallery of options on the Add New Wallpaper screen.

You may include pictures, people, a photo shuffle, emoticons, and a weather backdrop at the top.

The categories are Photo Shuffle, Suggested Photos, and Featured.

To set one of the wallpaper choices as your lock screen, tap it.

Swipe left or right to explore color filters with patterns and typography for background options.

In addition, you may touch widgets, the clock, and the calendar to adjust them.

When you tap one, a menu with many customization choices will appear.

When done, tap Add at the top, and then choose:

- Decide whether you want the wallpaper to appear on the home screen and lock screen.
- Then choose Set as Wallpaper Pair.
- Make more adjustments to the home screen.
- Click on Customize home screen. Then, you may blur it, among other things.

EDIT A LOCK SCREEN

Swipe to the lock screen you wish to alter, touch and hold your lock screen until the Customise button appears, and then press it. You may now make changes and save the document.

DELETE A LOCK SCREEN

By tapping and holding a lock screen until the Customize button appears, you may remove it. To erase a lock screen, slide to it, then swipe up and hit the trash icon on the screen.

SWAP OUT THE LOCKS ON THE DOORS

You may make many different personalized lock screens and swap between them as needed. Swipe to the lock screen you want to use and press it after touching and holding the lock screen until the Customize button appears.

HOW TO ADD A PHOTO TO YOUR IPHONE LOCK SCREEN

You may upload a particular picture from your photo library or choose a recommended photo when customizing your lock screen. Adding a Depth Effect to your background image or setting your lock screen to randomly choose photographs are more options.

The Customise button will display when you touch and hold the lock screen; hit the +.

Select Photos or Photo Shuffle by tapping + at the bottom of the screen.

If you want to create a multilayered look using Photos:

- At the bottom right, tap (...), then choose Depth Effect.
- Widget-enhanced backgrounds do not support layering.
- If the topic is too high, too low, or obstructs the clock, it could not be accessible.
- Only photographs of people, animals, or the sky may use layering.
- By pinching to zoom or dragging with two fingers, you may reposition your picture.
- Swipe left or right to experiment with color filters that include patterns and typography.
- By touching the grid symbol after selecting Photo Shuffle, you may get a preview of the photographs.

IPHONE 14 USER GUIDE

- Tap to choose the shuffle frequency (...).
- Select a choice after Shuffle Frequency.

Reminder: In the Photos app, pick Library > select a photo > press the Share button > choose Use as Wallpaper > tap Done > decide whether to display it on the home screen or lock screen.

HOW TO ADD WIDGETS TO YOUR IPHONE LOCK SCREEN

You may add widgets to your lock screens, such as those that display the current temperature, battery life, upcoming events in the calendar, and so on.

Touch the lock screen while holding down the button until it says "Customize."

Select Customize.

In order to add widgets to your lock screen, tap the box underneath the time.

You may add widgets by tapping or dragging them.

Simple!

HOW TO ADD A FOCUS TO YOUR IPHONE LOCK SCREEN

Your ability to focus may help you cut down on distractions. You may either build a new Focus or modify one of the pre-existing Focus choices, such as Work, Personal, or Sleep. For example, you may temporarily turn off all alerts by using Focus. The Focus may then be configured directly on your lock screen. By sliding to the associated lock screen when a Focus is attached to your lock screen, you may activate it.

You must first create a Focus:

Touch the lock screen while holding down the button until it says "Customize."

To see the Focus settings, tap Focus towards the bottom (for example, Sleep or Work).

After choosing a Focus, press X.

HOW TO CUSTOMIZE YOUR IPHONE HOME SCREEN AESTHETIC

Apple used to prevent its customers from personalizing the home screen. However, the business is now considerably more tolerant. An iPhone's aesthetic may be totally customized, for example, by adding widgets and replacing the app icons on the device's home screen with the one of your choosing. You may even add a video clip to your lock screen if you want to spice things up a lot. It's easy to complete.

OTHER IOS 16 TRICKS

MAKE NOTIFICATIONS INTO A LIST AGAIN

For the home screen, one more modification! You may use a list instead of the stacked notifications if you like them. Select List by going to Settings, Notifications, and Display As.

Audit what apps are accessing your data

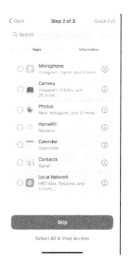

In order to enable individuals in perilous or abusive circumstances to know who may access their data and take appropriate action, Apple implemented a much-needed safety feature. Everyone should check out this option called Safety Check, which serves as quick security and privacy assessment. It will display the individuals and applications that have access to your microphone, camera, and location.

Access Safety Check by going to Settings Privacy & Security (near the bottom of the page). Go through the recommended check by tapping Manage Sharing & Access, then tapping Continue. The app list is organized to make it simpler to view and restrict access to applications. It is comparable to current Privacy & Security options.

MAKE YOUR KEYBOARD VIBRATE, QUIETLY

When using the on-screen keyboard, get a delightful clicking sensation. By switching the on Haptic option to green, go to Settings Sounds & Haptics Keyboard Feedback.

(If your keyboard produces audible clicking noises as you type, this is not how you must live. Toggle the Sound option off there so that it is gray.)

MAKE VOICE CALLS TO END THEM.

You may now end a phone conversation without pressing a button if you want to converse while knitting or walking. Access Settings, Siri & Search, and Call. Hang up and turn on the switch. The next time you're on a phone or FaceTime call, sternly command Siri to end the conversation. Excellent for claiming the last word.

DICTATE EMOJI

Emojis may now be included in essential emails and texts that you dictate to your employer without having to edit them, afterwards. Say the desired emoji out loud, then the phrase "emoji." For instance: "Sad face emoji since the presentation won't be done on time." (This is compatible with iPhone XS and later.)

GET RID OF THE SEARCH BUTTON

You no longer need to drag your finger down from the top of the screen to search for an app or contact since there is a little Search button at the bottom of the home screen. Toggle off the Show on Home Screen option under Search in Settings Home Screen if you don't like it.

CREATE REAL-TIME CAPTIONS

Even if the new function isn't as effective as genuine captions, it may still be helpful. Live captions will display a black box at the bottom of your screen that makes an effort to create captions for any open apps, including TikTok, podcasts, and your own videos.

HIDDEN IOS 16 FEATURES

Not every update in iOS 16 is immediately apparent.

The primary new features include a customizable lock screen, the ability to modify and remove text messages, and the simple removal of people, animals, or objects from photographs.

However, iOS 16 includes far more than simply those features. You'll be pleased to learn that the iPhone includes a number of new hidden functions that may significantly affect how you use it..

Here's what you need to know if you want to learn more about some of the hidden treasures that iOS 16 has to offer, such as password-protected picture albums and readily available Wi-Fi passwords.

PAIR NINTENDO SWITCH JOY-CONS WITH YOUR IPHONE

Apple has long enabled users to connect third-party controllers to your iPhone and iPad in order to play mobile video games like the Apple Arcade collection, Minecraft, and Call of Duty more comfortably. Examples of these controllers are the PS5 Sony DualSense and Xbox Core Controller. You may now increase that list by one more console controller.

You may now connect your Joy-Con controllers to an iPhone or iPad running iOS 16 if you possess a Nintendo Switch. Hold down the Joy-little Con's black pairing button initially until the green lights begin to flicker. The device is in pairing mode if this is the case. Go to Settings > Bluetooth on your iPhone after that and choose the Joy-Con from the list. Use the second Joy-Con to repeat this process.

RECENTLY DELETED AND HIDDEN ALBUMS ARE NOW PASSWORD-PROTECTED.

The Photos app's Hidden album is obviously not hidden since anybody can easily locate it. It is thus impossible to effectively conceal private images and videos. The Hidden album can be made "invisible" by Apple, but anybody with access to your phone might make it accessible once again and see everything within.

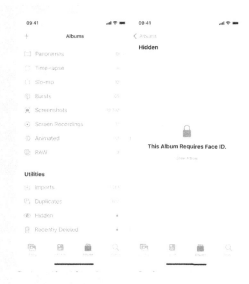

You may now lock the Hidden album thanks to iOS 16. Actually, there is nothing you need to do to turn on this functionality. Launch the Photos app and go to the Albums tab at the bottom of the screen to see it. The Hidden and Recently Deleted albums have a little lock next to them if you scroll down. You'll need to enter your password or Face ID to see the contents of those albums.

VIEW AND EXCHANGE STORED WIFI PASSWORDS.

Since a while ago, Apple has permitted iOS users to exchange Wi-Fi passwords, but only when two Apple devices are close to one another. Additionally, you couldn't just get the password from your settings if that function didn't operate automatically. Additionally, you had to remember the password if you wished to share a stored Wi-Fi password with someone else, such as an Android user or someone using a PC. Until now.

Tap the little information icon to the right of the network for which you want the password in the Settings' Wi-Fi section. Tap the Password area, then use Face ID or your passcode to display the network password. The password may then be shared by tapping Copy once it has been copied to your clipboard.

LOCATE AND ELIMINATE DUPLICATE IMAGES AND VIDEOS.

Perhaps you downloaded a film more than once or saved the same picture more than once, leaving copies all throughout your photo album. If you have enough capacity, it may not be an issue, but if you're out of room, iOS 16 makes it simple to eliminate every identical file.

You ought to notice a new Duplicates album under Utilities under Photos > Albums. Apple searches through your whole picture library and displays every image or video that has been stored more than once in a certain album. From there, you can choose to remove any duplicates or just hit Merge to preserve the best picture and its associated data while deleting the rest.

You may use Select > Select All > Merge to eliminate all of the images and videos that Apple deems duplicates at once, however you should definitely go over each set of duplicates to be sure they are exact copies and not just similar images.

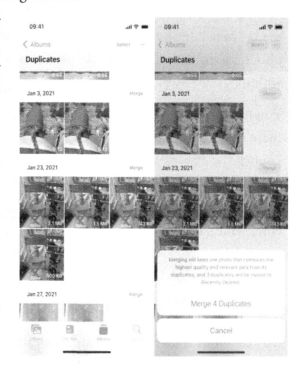

PICTURE AND VIDEO ADJUSTMENTS THAT ARE COPIED AND PASTED

You'll be delighted to know that you can now copy and paste modifications, such as saturation, contrast, and brightness, across photographs if you utilize the editor tool in the Photos app. You may paste the exact same modifications to any other image or video in your camera roll once you've finished editing one and are satisfied with the result.

Launch the Photos app, then open an altered photo in full-screen mode to do this. Next, choose Copy Edits from the three-dot menu in the top-right corner. Only if the image has been modified in Photos and not a third-party editor will you see this choice. After that, go to the image you wish to copy these adjustments to and choose Paste Edits by tapping the three-dot option. You should notice the picture adjustments appear after a little while.

SAFARI LETS YOU "PIN" YOUR PREFERRED TABS.

Safari only allows 500 open tabs, so if you're getting close to that number, it could be challenging to locate the precise tab you're searching for. Although you may browse indefinitely, there is now a quicker method to get the precise tab you need.

Pressing down on an open Safari tab now allows you to choose Pin Tab. This will pin that tab to the top of Safari where it will always be visible as a small tab preview that you may touch to see.

A tab will jump to the top of your grid of open tabs if you hold down the button and unpin it.

CHAPTER SIXTEEN
HEALTH APP WITH IOS 16

With the release of iOS 16, Apple included a new function in the Health app called Medication Management, which helps users keep track of their meds and never miss a dosage. The function complements the Apple Watch's new Medications app and may be used with both prescription and over-the-counter medications and supplements.

This section will walk you through the ins and outs of the Health app's new features in iOS 16, including the medication monitoring function.

MONITORING MEDICATIONS

The Health app now has a new area called "Medications," which can be accessed in two ways: by searching for "Medications" in the app's search box, or by browsing it in the Browse tab and scrolling down to it.

You can keep track of every drug, vitamin, and herb remedy you take with the help of the Medications app, which will then send you reminders when it's time to take your pills and analyze your data for any drug interactions.

In the Health app, selecting the Add Drug button is all that is required to add a new medication. There, you may search for a certain medication or vitamin by typing its name or scanning its label with your iPhone's camera. Scan-based addition is convenient, but it doesn't always capture all the data, so you may have to choose the dosage and form by hand (pill, spray, etc).

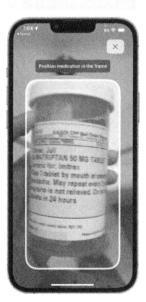 

IPHONE 14 USER GUIDE

It's up to you from there how often you want to take medicine. You may schedule your prescription doses to be taken at any time of day, on any day of the week, or on an as-needed basis.

Medicines may be made to look like the actual pill you take, down to the color and form.

The Health app has a place for you to keep track of the vitamins and prescriptions you take.

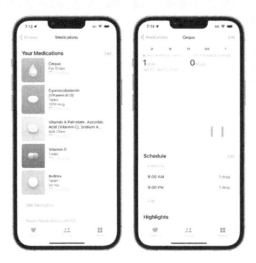

Unless you change the option, medication reminders will be sent at the time you choose since they are considered time-sensitive alerts.

All drugs with a regular time and/or day have a reminder sent to you automatically.

MEDICATION LOGGING

When a reminder alert appears, tapping it will open a window where you may choose "Taken" or "Skipped" for a drug. If many prescriptions are due at the same time, you may utilize the "Mark All as Taken" function. The Health app's medication list also allows you to record medications by checking the appropriate box.

IPHONE 14 USER GUIDE

You may monitor your drug intake using the Health app over time to make sure you're taking your pills as prescribed. Information is presented daily, weekly, monthly, semiannual, and annual.

In addition to the Health summary, Apple will provide Medications "Highlight" that details how frequently a medicine has been taken over the last 28 days.

APPLE WATCH MEDICATION APP

WatchOS 9 has a Medications app specifically designed to make tracking your medications easier. The software allows you to record each dose taken separately or all at once by tapping "Log All as Taken." The "Skipped" option is used when a dosage has been missed.

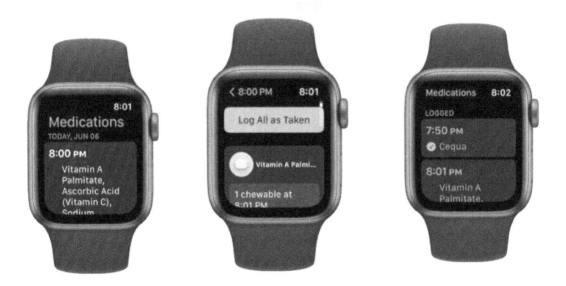

All medication management should be done in the Health app; the Apple Watch app is just used to check off when pills have been taken.

MEDICATION INTERACTIONS

The Health app performs frequent checks to ensure that the drugs you are taking do not interact with one another, preventing you from accidentally taking a harmful dose of more than one drug.

ETHAN COPSON

MEDICATION SIDE EFFECTS AND OTHER INFO

Side effects information, if available, may be seen by tapping the "Side Effects" button after selecting a drug from your Health app's list.

Elsevier is a Netherlands-based medical content provider that publishes a variety of resources, including the Annual ScienceDirect book series on pharmacological side effects, but not for every prescription.

EXPORT MEDICATIONS

You may generate a basic PDF of your medication list by tapping the "Export Medications List PDF" button in the Health app's Medications section. A doctor would benefit from seeing this list.

SLEEP FEATURES

The new features included in watchOS 9 and iOS 16 are especially useful for users who wear their Apple Watch to bed in order to keep track of their sleep patterns. The Apple Watch records your sleep and classifies it as Awake, REM, Core (light), and Deep so you can see how well you slept.

When you go into bed, how long it takes you to fall asleep, how frequently you wake up, and how much time you spend in REM, Core, and Deep sleep can all be monitored with an Apple Watch.

Based on these criteria, Apple classifies sleep as follows:

Awake - It's not uncommon to have periods of wakefulness throughout a night of sleep. On rare occasions, humans will awaken. It's also likely that you'll go back to sleep after awakening and not even realize that you were awake.

Rapid Eye Movement (REM)- sleep has been hypothesized to contribute to memory consolidation and learning retention. At this point, you're completely at ease, and your eyes are darting quickly from side to side. Most of your dreams also occur at this time.

Core - This stage, also known as light sleep, is believed to be of equal importance to others. It's common for this to account for the bulk of your nightly sleep. Cognitively relevant brain waves are generated at this time.

Deep - This phase of sleep, often called slow wave sleep after the associated brain wave patterns, is crucial for the body's recovery and hormone production. This phase of sleep occurs more often during the first half of a normal night's sleep.

The Apple Watch logs your sleep habits in the Health app, under the "Sleep" tab. Your sleep data is no longer limited to a weekly overview; instead, you may see it by the day and examine your nightly patterns in more detail.

The functioning of Sleep has not changed other than the additional sleep types. The same Bedtime and Wake Up functions are used, and they are linked to the Wind Down and Sleep Focus settings.

HEALTH SHARING INVITATIONS

With iOS 16, you can easily maintain tabs on the medical records of elderly relatives or young children by inviting them to share their data with you. After accepting your invitation, the invited party has the option of choosing which pieces of information to share with you.

CHAPTER SEVENTEEN
SETTING UP FITNESS APP WITH IOS 16

You may now use the Apple Fitness app to monitor your activity levels without an Apple Watch, thanks to iOS 16 (public beta available since July, final version due in September). In this article, we will teach you how to set up the Fitness app on your iPhone and begin monitoring your health immediately.

HOW TO USE THE APPLE FITNESS TRACKER APP ON AN IPHONE

Until recently, the Apple Fitness app could only be used in combination with an iPhone and an Apple Watch. In previous iOS it was necessary to use third-party software for even the most fundamental functions, such as counting steps. The Fitness app formerly required an Apple Watch to function, however with the release of iOS 16 (beta or final), it is now available to everyone.

If you don't have an Apple Watch, here's how to get started using the Fitness app on your iPhone:

You may launch the Apple Fitness app by clicking its icon.

When you initially launch the app, you'll get a brief welcome message. Simply press the Next button. You probably won't see the first setup screens or the welcome message if you've already started the app before.

When you initially launch the app, you'll be asked to authenticate your health data. If you've already entered this information, it may be correct; if not, simply tap the portion that needs to be corrected and enter the correct information.

When everything checks out, click the Next button.

IPHONE 14 USER GUIDE

Next, Determine your Daily move goal. The initial sum will be somewhat little, and it will be possible to increase it if needed. You may adjust your calorie intake by tapping the Lightly, Moderately, or Highly buttons, or the Plus or Minus symbols right next to the Calorie count. Simply press the Next button when you're ready to go forward.

Finally, Whether you want to get alerts, the app will ask you if you want to. Follow the prompts by clicking the Next button.

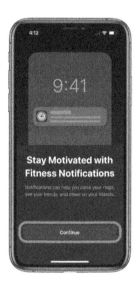

Next, decide whether you want to get alerts by tapping the Allow or Don't Allow button on the pop-up box.

Now, Tap the little Get Started in the Trends box on the main Summary page to learn more.

Learn more about Activity Trends and go on by selecting Next.

If you want to ask friends or family to share their Fitness activities with you, you may do so by tapping the Sharing button in the bottom right of the Summary page.

IPHONE 14 USER GUIDE

Next, Select the little contact symbol located at the top right.

Simply choose the Plus symbol, located at the top right, and then choose the people you'd want to add.

Contacts may be accessed here by doing a search. You may either scroll through the list of suggested contacts or use the To box to do a direct search for a specific contact. Select the desired recipient(s), and then click the Send button.

The last thing you need to know about the app's setup is the little Account icon in the top right. It will resemble the photo you used to create your Apple ID. If you tap this button, the main settings menu will appear.

You may modify any of the options and details you selected during the app's initial setup here.

Everything you need to get going should now be in place. Your activity levels, including the number of steps taken, the number of flights of stairs climbed, and the number of calories burnt, will all be recorded by the Fitness app.

ISSUES & TROUBLESHOOTING

Using the Fitness app without an Apple Watch has been smooth sailing for us, but some on the iPhone Life team have run into problems. If you discover that the app isn't recording your steps and activity as intended, here are a few things to check.

UPDATE SOFTWARE

Check to see whether there are any outstanding iOS 16 system software upgrades and complete them. Whether you have an Apple Watch and Fitness isn't functioning as expected, check to verify if you're on the most recent version of Watch OS 9.

CHECK MOTION CALIBRATION & DISTANCE SETTINGS

If you turn off an essential option, Fitness will not be able to keep track of your activities. What you're looking for may be found as follows: To start, go to the Settings menu.

Then, choose "Privacy & Security" from the list that appears.

The function you want, is situated at the very top of this page: Location Services

Make sure the foremost Location Services toggle is turned on, then scroll all the way to the bottom and choose System Services.

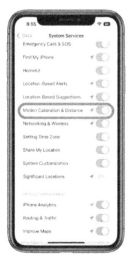

There are several other location-related toggles on this page, but the one we're interested in for Apple Fitness is the one under Motion Calibration & Distance. Make sure this switch is turned on (green); if not, step and activity tracking won't function properly or at all.

You should be able to use Apple Fitness to monitor your activity after making sure your system software has been updated and that Motion Calibration & Distance is enabled under Location Services.

CHAPTER EIGHTEEN
TRICKS TO BOOST BATTERY LIFE

Tests on the battery life of the iPhone 14 have already produced results. Though it can't quite top its predecessor, the iPhone 14 Pro Max is the greatest. Both the iPhone 14 Pro and the iPhone 14 has great battery life that should meet the demands of the majority of users. Although the iPhone 14 Plus has not yet been released, it should likewise have fantastic battery life.

Nevertheless, there are several things you can do to guarantee the optimum battery life for your iPhone 14 Pro. All iPhone 14 models can benefit from the following advice, while certain ideas can only be used with the iPhone 14 Pro and iPhone 14 Pro Max.

TURN OFF THE ALWAYS-ON DISPLAY ON IPHONE 14 PRO AND PRO MAX

We'll start with a function that the iPhone 14 Pros have just. The phone's capability for always-on displays means that the display is virtually always on. The brightness may be reduced in iOS 16 and the refresh rate will decrease to 1Hz. But the always-on display continues to consume power. Or put another way, it affects battery life.

To turn off Always On, go to Settings, Display & Brightness, and deactivate it.

A further unanticipated drawback of an always-on display exists. The user experience has been criticized by a number of reviewers. It is very bright on the darkened screen.

We've learned through years of iPhone usage that the screen only illuminates when you get a notice. As a result, if the screen is on, you may be more inclined to pick up the iPhone 14 Pro and Pro Max.

It's a reflex to unlock the device to check the alerts, only to find that there are none. Additionally, as you would have anticipated, you are reducing battery life.

If you must use the always-on display, be aware of all the circumstances in which the screen totally shuts off so that you may take advantage of them to save battery life.

TURN OFF KEYBOARD HAPTICS IN IOS 16

A popular new feature in iOS 16 is support for haptic feedback on your iPhone's keyboard. The keyboard now vibrates in addition to creating noises to let you know when a key has been pressed.

However, such combination will use energy. The iPhone must provide vibrations and play noises. Additionally, Apple clearly informs consumers on a support page that "turning on keyboard haptics can decrease the battery life of your iPhone."

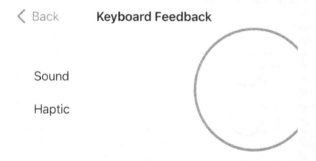

On the iPhone 14 and iPhone 14 Pro models, you should disable the Sound and Haptic to extend battery life. Go to Sounds & Haptics and Keyboard Feedback in the same Settings app. Toggles for both settings are located there. Make careful to disconnect them both.

ENABLE THE IPHONE'S AUTO-BRIGHTNESS FEATURE

The greatest energy is used by the OLED display on your iPhone 14 or iPhone 14 Pro. It consumes energy to show colors other than black since it employs OLED technology. The more power your battery can provide for the experience, the brighter it becomes.

Make sure Auto-Brightness is enabled to maximize battery life. Accessibility, Display & Text Size are where you may locate the setting. The iPhone's ambient light sensor(s) will automatically lower the screen brightness when you activate it. Not only is this good for the eyes, but it also conserves battery life.

You may be able to extend the life of your battery by using Dark Mode constantly. An OLED panel uses no energy while displaying black.

LIMIT PROMOTION FRAME RATE

It's possible that you'll be tempted to control the Motion settings while you're in the Accessibility menu. You may also activate the Limit Frame Rate option in that menu. This additional battery-saving tip is exclusive to the iPhone 14 Pro and Pro Max.

The ProMotion function will be rendered useless since the feature will limit the maximum refresh rate on Pros to 60Hz. Remember that while the option is enabled, the screen refresh rate will still vary between 1Hz and 60Hz. You may check that by turning on Always On. If the refresh rate of the screen cannot be reduced to 1Hz, this functionality will not function.

By keeping the screen from increasing to 120Hz, you'll save battery life. Although ProMotion is a fantastic feature for the iPhone, it is not as revolutionary as 120Hz on Android. Some folks may not even notice or give a damn about the changes between 60Hz and 120Hz on iPhones. They have the option to activate the Limit Frame Rate function.

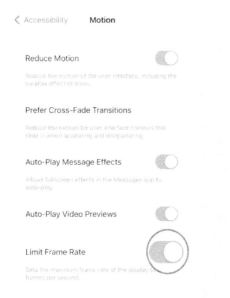

REMEMBER BATTERY HEALTH

The aforementioned sneaky methods won't work miracles. It is likely that battery life may increase by a few minutes but not by many hours. Nevertheless, any advancement would be welcome news for those who have battery life concerns.

These iPhone owners need to keep this in mind to maintain the battery health of their devices. The Battery Health options, found under the Battery menu, must have Optimized Battery Charging activated.

The function slows down battery aging by studying your daily routine. Only when you need it will the iPhone 14 or iPhone 14 Pro complete charging beyond 80%. For instance, the phone will finish charging before your morning alarm goes off.

Your iPhone should provide more than simply exceptional battery life. This little adjustment will help you obtain the long battery life you desire from it.

Regarding that battery worry, you may always turn on Low Power Mode to save energy in an emergency.

IOS 16 WILL STOP CHARGING WHEN THE IPHONE IS HOT

A new built-in feature of iOS 16 that is available on all iPhone models is another battery safety function that will enhance battery health and lengthen battery life.

The following message indicates that your iPhone 14 model is too hot to recharge if you see it:

Holding Charges. When the iPhone reaches its usual temperature, charging will start again.

The functionality may be used without anything in the Settings app being enabled. However, you may want to take some action, such as taking the iPhone out of the sun. Alternately, avoid using energy-hungry applications while the phone is charging.

GET THE RIGHT BATTERY CHARGER FOR THE IPHONE 14

Battery charging speed is one of the iPhone 14 Pro speculations that didn't come true. Instead of the predicted 30W, the claimed maximum charging rate for the iPhone 14 models is still 20W. That will give you a 50% charge on any iPhone made after the iPhone 8/X in 35 minutes.

Some Android smartphones allow for charging at 100W or more. These speeds, however, may harm the battery's health. In turn, they will deplete batteries more quickly than phones with slower charging times.

Although Apple doesn't promote it, the iPhone 14 Pro does offer somewhat quicker charging rates. With earlier iPhone models, we saw this, and Apple is covertly carrying it out once again. Some models could get close to the 30W threshold even if the iPhone 14 and 14 Pro might not. According to user tests, the iPhone 14 Pro Max can charge up to 29W when using a bigger USB-C port.

For the iPhone 14 and iPhone 14 Pro Max, the typical power consumption seems to be approximately 25W and 27W, respectively.

To somewhat speed up charging, you may choose to utilize a bigger model rather than a 20W charger. Any standard USB-C charger, including the power brick for your MacBook, may be used to recharge the iPhone.

Although you won't reach a steady charging speed of 25–27W, the bigger charger may shorten emergency charging times and "eliminate" battery anxiety. All of this without drastically reducing battery health.

CHARGING YOUR IPHONE 14

Although they don't come with them, the newest iPhones do allow wireless MagSafe charging. These are the quickest methods for charging the iPhone 12, 13, and 14, whether or not you utilize a cable.

In an effort to save packaging waste, Apple no longer supplies wall adapters in every box as of the iPhone 12. (and make some cash on accessories). Its most recent devices accept

Apple's MagSafe magnetic charging, including the iPhone 14 series unveiled this week. The information you need to know about charging your iPhone 12, iPhone 13, or iPhone 14 is provided here, along with any possible shopping list items.

WHAT COMES WITH THE IPHONE 14?

The USB-C to the Lightning connection that comes with the most recent iPhones is pretty much it. Those without Apple power adapters will need a USB-C power adapter to charge right out of the box.

Additionally, the latest iPhones don't come with EarPods, so you'll need to buy your own to listen to podcasts and music. In addition to our choices for the finest wireless headphones and those designed with runners in mind, Apple offers its own AirPods wireless earphones (and the second generation of AirPods Pro goes on sale next week).

According to Apple's explanation during its iPhone 12 announcement a few years back, leaving off the power adaptor makes the package smaller. The corporation can now supply stores more quickly and save annual carbon emissions by 2 million metric tons since 70% more gadgets can fit on a shipping pallet on their route to consumers.

HOW DO I CHARGE THE IPHONE 14?

Apple has not completely switched the iPhone to USB-C, which normally offers quicker charging rates, unlike most new Android phones, therefore the iPhone 14 still retains the standard Lightning charge connector. This means that you may charge with your current Lightning connection and a standard USB-A wall adapter. However, you may also connect your iPhone directly to a USB-C port-equipped item, like a lamp or light switch, using the USB-C to the Lightning connection that comes with your iPhone.

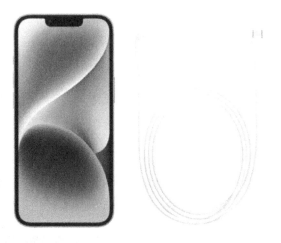

The most recent iPhones also work with Qi wireless charging mats. Apple's built-in MagSafe magnetic power connection, which supports rapid charging, is its primary emphasis for charging.

WHAT IS MAGSAFE?

Apple has referred to the charging cable connections on their laptops as MagSafe for many years. They "snapped" into the magnetic MacBook charging ports and "snapped" out if disturbed, for example, to prevent a Mac laptop from falling to the ground. They vanished a few years ago when

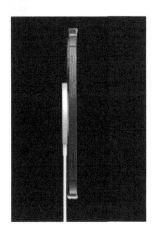

Apple switched its MacBook series to USB-C for data transmission and charging, but they made a comeback last autumn in the form of the "MagSafe 3"-based M1 Pro/M1 Max MacBooks.

Recent iPhones from Apple come with a similar feature in the shape of a magnetic "hockey puck" disc that clips to the back of the device and resembles a large Apple Watch charger. This USB-C cable with the MagSafe connection fits into a power outlet and charges devices at a 15W rate.

WHAT'S THE FASTEST WAY TO CHARGE THE IPHONE 14?

With the exception of the iPhone 11 Pro and Pro Max, which came with an 18W USB-C power converter, Apple has always sent its iPhones(Opens in a new window) with a 5W USB power adapter inside the box. Depending on the model, Apple's iPads also come with 10W, 12W, 18W, or 20W USB-C power adapters. Therefore, unless you started buying iPhones after the iPhone 12, you presumably still have one around.

Apple's smartphones, on the other hand, feature fast charging with the release of the iPhone 8, which enables a 50% charge in 30 minutes. However, in order for rapid charging to function, you must use an Apple USB-C-to-Lightning connection and one of the following adapters: an 18W, 29W, 30W, 61W, 87W, or 96W USB-C Power Converter from Apple; or an external USB-C power adapter that supports USB Power Delivery (USB-PD).

With Qi wireless charging, Apple iPhones 8 and later can be charged wirelessly at 7.5W using a compatible wireless charging pad. The Apple-branded MagSafe charging system will charge your phone at 15W, which is even quicker than a Qi charger.

A Qi charger, on the other hand, has a tendency to slide out of position and leave you scurrying for juice at the last minute. MagSafe, however, uses a magnetic snap to verify that your phone is truly charging.

Therefore, MagSafe will provide the quickest, most dependable charging times provided you have a 5W charging adapter, a Qi wireless charging station, and that connection. The issue is whether the additional $39 for the MagSafe connection is worthwhile. The answer may be yes if you always require a charge. Others may be able to get by with an old power adapter or charging pad, and $40 might be better spent on a tough case to safeguard your expensive new smartphone.

WHAT DO I NEED TO BUY?

You don't need to purchase anything if you already own an iPhone or iPad with a Lightning cord. You're set to go when you put your existing power adapter into the wall and your iPhone 14 with the Lightning connector.

The MagSafe charger and 1m USB-C integrated cable cost an extra $39 if you want to use the "snap" effect that secures and charges your phone (you have other options, too). For $129, Apple also offers a foldable MagSafe Duo Charger with an Apple Watch dock.

The included Lightning-to-USB-C cable and the MagSafe charger may both be used with a separate 20-watt USB-C power converter(Opens in a new window) for quicker charging than Apple's 5W power adapter. It was formerly $29, but is now just $19.

Although the iPhone 14 comes with a Lightning-to-USB-C connection, Apple also offers longer Lightning-to-USB-C cables that can be purchased for $19 or $29 for 1m(Opens in a new window) or 2m(Opens in a new window) lengths. On Amazon, you may also locate less expensive options.

DOES MAGSAFE WORK WITH A CASE?

A MagSafe-compatible accessory is required if you wish to combine magnetic charging and a cover to protect your iPhone 14. Naturally, Apple offers several MagSafe-compatible cases and a magnetic wallet. There are other MagSafe devices made by other companies, like Belkin and Otterbox.

The MagSafe charger should work with non-MagSafe cases without a magnet since it is Qi wireless compatible, but it won't attach to the case.

DOES MAGSAFE WORK WITH OLDER IPHONE MODELS?

The MagSafe charger is theoretically compatible with previous smartphones that enable wireless charging, as well as with variants of AirPods that have a wireless charging cover, even though it was designed to operate with the iPhone 12 and higher (see officially compatible models below).

However, you won't be able to "snap" your iPhone into the optimal charging position since these earlier phone models lack the magnetic component on the back of the device. That is only supported by the iPhone 12, 13, and 14 models. You can easily set any other phone on the MagSafe connection (even certain Android phones (Opens in a new window)), but it won't lock into place.

Supported iPhone Models
- iPhone 14
- iPhone 14 Pro
- iPhone 14 Plus
- iPhone 14 Pro Max
- iPhone 13

- iPhone 13 Pro
- iPhone 13 mini
- iPhone 13 Pro Max
- iPhone 12
- iPhone 12 Pro
- iPhone 12 mini
- iPhone 12 Pro Max
- iPhone 11
- iPhone 11 Pro
- iPhone 11 Pro Max
- iPhone SE (2nd generation)
- iPhone XS Max
- iPhone XS
- iPhone X
- iPhone XR
- iPhone 8
- iPhone 8 Plus

Supported AirPods Models

- Wireless Charging Case for AirPods
- AirPods Pro (including 2nd generation (Opens in a new window))
- AirPods with Wireless Charging Case (2nd generation)
- AirPods (3rd generation)

FAST CHARGERS FOR APPLE IPHONE 14 SERIES

With a host of advancements and groundbreaking capabilities like Emergency SOS via Satellite, the much anticipated iPhone 14 series has finally arrived, no charger included. This implies that if you want to use your brand-new iPhone for more than a few hours, you'll need to buy a charger separately. But don't worry, we've selected a number of fast chargers that you can purchase for the iPhone 14 series.

We have everything covered, from portable single-port micro chargers to multi-port travel chargers that can even recharge your laptop. The best chargers for the iPhone 14 series are listed below. iPhone 14, iPhone 14 Plus, iPhone 14 Pro, and iPhone 14 Pro Max can all be quickly recharged using the included chargers.

Here are the best chargers for iPhone 14 and iPhone 14 Pro series.

1. APPLE 20W USB-C POWER ADAPTER

Since you purchased your phone from Apple, you may choose their original charger. The Apple 20W wall adapter is ideal for charging the iPhone 14 series of mobile devices. It is a straightforward charging brick without any fancy bells and whistles, featuring a USB-C connector on one end.

This Apple adaptor will work well for you if all you need is a straightforward, no-frills charger for your iPhone 14. Many customers may want to continue using a first-party charger for compatibility and safety concerns. You should use this charger if you fall into such category.

With a maximum wattage of 20W, it can charge your iPhone 14 and iPhone 14 Pro devices.

Additionally, it is among the most reasonably priced charging bricks, which is a crucial factor for the majority of individuals. You may use the same adapter to power your iPhone 14 series phone if you want to charge it using MagSafe. There are more than 150,000 reviews of the Apple 20W charger on Amazon, most of which are encouraging, so buyers can feel confident. However, if you use various gadgets, keep an eye out for more adaptable choices.

2. SPIGEN 27W USB-C FAST CHARGER

Here is a GaN charger that is both very compact and portable for carrying about and can charge your iPhone 14 quickly. Because of the greater wattage, you may use this converter to charge your iPad and other Android devices as well.

Its small size makes this Spigen USB-C charger convenient to carry in your pocket. It has a larger output wattage of 27W, however, so your gadgets will charge more quickly. The iPhone 14 Pro and Pro Max can charge up to 23W, while the iPhone 14 and iPhone 14 Plus can charge at a maximum of 20W. Therefore, other gadgets like an iPad or an Android smartphone benefit from the increased watts.

This adapter's compact factor is its strongest point, however, one drawback is that its prongs are not collapsible. But bear in mind that some consumers have expressed concerns about the adapter becoming heated after extended use. There is no need to worry since it shouldn't slow down the charging rates. If you have other devices that can charge at a greater wattage in addition to your iPhone, the Spigen charger will be useful to you. Additionally, it offers a longer warranty than other options at 24 months.

3. ANKER NANO 3 USB-C 30W CHARGER

In terms of physical size, the Anker Nano 3 is a 30W USB-C charger that resembles Spigen's adaptor. With this charger, you do, however, get a bonus in the shape of a folding plug. This makes it simpler to transport, particularly while traveling, in your pocket or bag.

This recently announced charger from Anker offers a single USB-C connector and a slightly greater output of 30W than the Spigen adaptor. So, neither the speed of charging nor the length of time needed to fully charge the phone will alter. Of course, this is useful for charging 30W-capable electronics. The connectors on this adaptor are foldable when not in use, which makes it different from others.

Foldable prongs are a godsend if you are someone who stuffs their charger in their jeans pockets or if you travel often and keep your charger in a bag since they make the charger take up less room. Oh, did we forget to add that the adapter matches the new iPhone 14 Pro's purple color by being purple as well? Talk about giving a charger a special and enjoyable touch!

4. UGREEN NEXODE MINI 45W CHARGER

The two USB-C ports inside Ugreen's dual-port PD charger provide a combined output of 45W. This makes charging two smartphones at once really handy. One port may be used to charge your iPhone 14 while the second, faster port charges your Apple Watch simultaneously.

You may choose to charge your iPhone 14 or iPhone 14 Pro device together with a different accessory, such as an AirPods or an Apple Watch. A dual-port adaptor might be useful in this situation. When both ports are utilized at once, this Ugreen GaN adapter's 45W output is split into 25W and 20W.

This output has the capacity to simultaneously fast-charge your Apple Watch and your iPhone 14. This may be a huge benefit if you're traveling. If you're just utilizing one USB-C port, the 45W output also has the benefit of being able to charge a laptop like the MacBook Air. It's the ideal travel companion thanks to the folding plug and convenient form factor.

5. ANKER NANO II 65W CHARGER

The Anker Nano II is a multi-port iPhone 14 series charger that can simultaneously charge your iPhone and laptop or Mac. Thanks to a folding plug design, it is extremely transportable and simple to use.

The Nano II from Anker is a portable charger that you may use while traveling for your laptop and phone. The Anker Nano II is a fantastic charger to use with the iPhone 14 and a laptop such as MacBook Pro since it can charge devices at up to 65W. Two USB Type-C connections on this iPhone 14 charging block allow for simultaneous rapid charging of two devices. A USB-A connector is also available for charging devices like the Apple Watch.

The folding prongs on this charger make it simpler to carry as they take up less room in your backpack. While having a third USB-C port rather than a USB-A port would be nicer, the more compact form size more than makes up for it. However, according to user reviews, this adapter reduces the output power from the secondary USB-C port while you're charging a device like a laptop, which might be a drawback.

KEEP YOUR NEW IPHONE CHARGED UP

Before you even grab the phone from the Apple shop, get an iPhone 14 charger because one is not included in the package! You should have no trouble charging your smartphone if you choose one of these USB-C chargers for the iPhone 14, iPhone 14 Plus, iPhone 14 Pro, or iPhone 14 Pro Max.

CHAPTER NINETEEN
WAYS TO SELL OR TRADE IN YOUR OLD IPHONE

Don't simply dump your old phone in a drawer when you upgrade, whether you're looking for a new iPhone 14 or want to go large with the iPhone 14 Pro. Sale of the iPhone! You may be surprised to learn how precious an old piece of glass is.

There are several businesses that will exchange your old iPhone for cash or shop credit. For trading in an outdated gadget, we evaluated a few well-known providers. All of these services support iPhones, and the majority also accept Samsung and Google smartphones.

TIPS TO GET A HIGH RESALE PRICE

Make sure to take care of your phone if you want to get the most money when you sell it. Purchase a sturdy case and think about a screen protector. The greatest approach to make sure you receive the most money when you sell your equipment is to maintain it looking brand new.

Buy an unlocked phone at all times. You will be free to swap carriers thanks to this, and you'll also earn more money when you decide to sell it. Any wireless network that supports its antenna may be used with an unlocked phone. You shouldn't be bound to a single carrier. Unless a carrier specifically states otherwise, a phone is often not unlocked, particularly if you purchase it on a payment plan.

Make sure you use iCloud to backup all of your data before rushing out to sell your old phone. To ensure that iCloud stores your text messages, which sometimes include pictures and videos that you haven't saved to your Camera Roll, be sure to tick the option to backup your Messages. If you have an Apple Watch, don't forget to unpair it and delete your phone's data as well.

1. BEST FOR PRISTINE IPHONES

Swappa

Swappa, a kind of eBay clone, aims to fix some of the issues with eBay, including its excessive seller fees, inadequate facilities for communicating with buyers, and abundance of low-quality products. Your iPhone must be in excellent condition, completely working, and undamaged in order to be sold here.

This is the place to sell your old phone if it satisfies Swappa's listing requirements and you're prepared to put in a little work. You'll need to put out a listing with images, just as you would on eBay. Make careful to remove the phone's cover and be upfront about its condition. When determining your selling price, don't forget to account for delivery.

A 128-GB iPhone 13 is now available for around $630. Depending on its storage size and condition, an iPhone 12 costs between $450 and $540.

2. BEST FOR RELIABLE CASH

Gazelle

In the realm of second-hand phones, Gazelle is an expert. The organization has the easiest procedure and has been purchasing phones since 2006.

You fill out an online form with information about your device, including its functionality, the carrier it is connected to, and any visible physical flaws. Following that, you'll get an offer based on your responses. If you decide to take Gazelle up on their offer, they'll give you a box with a mailing label, and you'll send the phone in for evaluation. You'll be paid once the business inspects your item and confirms that it is in the condition you claimed, which typically takes seven to ten days. Payment options include checks, PayPal, and Amazon gift cards.

You can purchase a 128-GB, factory-unlocked iPhone 13 for $448 in excellent condition. You can get a 64-GB, unlocked iPhone 12 for roughly $315. Keep an eye out for special deals from Gazelle since they sometimes coincide with the introduction of new devices.

3. BEST FOR A BROKEN IPHONE

uSell

uSell will still purchase your iPhone even if it has damage or if the battery can no longer be charged. uSell provides the most options for broken electronics among the send-it-in-style services. For an iPhone 11 that doesn't even turn on, we were given $80. You can receive at least $320 for a broken, factory-unlocked, 256-GB iPhone 11 Pro Max, which is better than nothing.

4. EASIEST WAY TO SELL

Apple's iPhone Trade-In

Although the pricing offered by Apple's Trade-In program isn't always the greatest, if you're purchasing a new iPhone, Apple will swiftly apply your credit to the new phone. Apple only pays top bucks for flawless phones.

An iPhone 12 in excellent condition now sells for $500, while an iPhone 11 may get up to $340. The alternative is to turn in your old phone for a discount on a new one. If you already own an iPhone 12, you can trade it in for a 128 GB iPhone 13 for $299 (a savings of 530 dollars) or an iPhone 13 Pro for $469 (a savings of 530 dollars).

Although it is not advisable to adopt Apple's iPhone upgrade program might help you save money if you're the kind of person who has to update to a new iPhone every year. It is less expensive than purchasing a new unlocked phone every year since you pay for your phone permanently every month but receive a new one after 12 months.

A FEW MORE OPTIONS

Best Buy: Trade-ins are available at Best Buy online or at participating shops. Damaged gadgets won't be accepted, and you'll only receive shop credit instead of cash. However, if you still need anything from Best Buy, going in-store is a reasonably hassle-free choice.

Carrier Trade-Ins: Particularly if you're switching to a new iPhone, your service provider certainly offers some form of buy-back program. Your handset will be purchased back by Verizon, AT&T, and T-Mobile, albeit often for far less than the other alternatives on this list.

eBay: There is eBay, always. Despite the fact that my recent personal experience has made me less enthusiastic about eBay—the number of buyer scams appears to significantly outnumber the number of reliable buyers—it is still one of the most widely used platforms for online sales. Even a unique form for selling your iPhone is available on the website.

Amazon and Others: Other device-specific exchanging websites, such as GadgetGone e.t.c. Although Walmart and Amazon both have programs for trading in devices, none of them gave a better bargain than the ones mentioned above.

Decluttr: Decluttr is a reliable solution to sell your phone quickly for cash. Although it never paid as much as other jobs, it was easy and quick. The WIRED Gear Team has had some negative experiences with the service, therefore we are hesitant to endorse it. Nevertheless, negative events may occur with any of these businesses. Decluttr could be worth a try if you're dissatisfied with the offers you're receiving from these other businesses.

Please take note that the latest introduction of the iPhone 14 series may have changed all of the prices stated.

CHAPTER TWENTY
YOUR IPHONE 14 PROTECTION

Every new iPhone owner has asked this question: Do you need to get a case and screen protector or can you just enjoy your iPhone with accessories?

If you're planning to buy a new iPhone, or if you already own one, you may have questioned whether you need a case and screen protector for it. Of course, you want to keep your iPhone well protected, but you don't want to waste money on these accessories if they aren't essential.

We'll look at the pros and cons of using an iPhone case and screen protector so you can decide what's best for you.

DO YOU NEED TO PROTECT YOUR IPHONE?

This question cannot be answered simply because it is dependent on how you handle and use your equipment. The iPhone is a costly gadget composed of premium components. It should withstand most common impacts, but if you drop it from a fair height onto a hard surface, it will still shatter.

You could appreciate using your iPhone without a cover or screen protector if you seldom dropped your prior phones and they didn't incur many scratches. This will enable you to enjoy the elegant appearance of a naked iPhone while also saving you some money.

On the other hand, a case and screen protector are unquestionably the best choice if you often drop or scratch your phones. Although you won't have the same experience as when using a bare iPhone, you won't always be concerned about losing it.

If you're prone to accidents, you may also want to get AppleCare+. Apple provides a service that functions similarly to insurance. You have the option of paying a monthly charge or two years' worth of service fees up in advance. Then, while it's still not free, you may get unintentional damage fixed at an Apple Store for a small fraction of the cost of standard repairs.

For individuals who dislike cases or screen protectors but are concerned about their iPhone breaking, this is a perfect substitute.

REASONS WHY YOU SHOULD USE A CASE AND SCREEN PROTECTOR

iPhone silicone covers piled up: If you're still unsure, consider these arguments in favor of getting a case and screen protector for your phone.

You're Accident Prone: Your iPhone is substantially better protected if you drop it unintentionally with a cover and screen protector. Because of this, it's far less likely that you will need to pay for an expensive repair or screen replacement. The case or screen protector will often only need to be replaced if they get damaged.

You Want a High Resale Value: Maintaining a high resale value for your iPhone is another incentive to secure it. Compared to other cellphones, the iPhone holds its value significantly better over time. However, if you use a case and screen protector to preserve your iPhone in pristine condition, its resale value will continue to be much greater.

A pricey repair is out of your price range: repairing an iPhone is costly. You must spend $279 if you possess an iPhone 13 Pro and need to replace the screen. Expect to spend up to $549 if further components need to be replaced. Spending $10 to $20 on a case might wind up saving you hundreds of dollars in the long run since you may want to eliminate any danger of having to pay that.

You want to customize your iPhone: There are many different ways to do this while using a cover. Apple gives you a few color options when you purchase an iPhone, but once you make a decision, you are locked into that hue for the duration of the life of the device, even if you no longer like it. To alter the look of your iPhone, you can always get a new cover; some websites even allow you to customize your phone case.

CONSIDERATIONS AGAINST THE USE OF A CASE AND SCREEN PROTECTOR

Naturally, not everyone wants or needs to spend money on a case and screen protector. Here are several justifications for not using an iPhone cover and screen protector.

You've Spent Enough Already

Your iPhone already costs a lot of money, but adding a cover and screen protector will increase the price even more. Sure, it might end up saving you money in the long run, but if you never drop your iPhone, you've wasted your money now and made no savings.

You want to use the iPhone as is and enjoy it

The iPhone is a high-end gadget with great design, built of premium materials. Why cover it with a case and screen protector if you purchased it because you enjoy the way it looks? Use your iPhone exactly as it is if you want to appreciate the design and feel the metal under your fingertips.

YOU WANT YOUR BETTER COOLING

The fact that iPhone covers often inhibit your phone from cooling down as effectively is another reason why you may want to avoid them. Stainless steel or aluminum are used to make iPhones. These two substances swiftly let off heat. However, it won't be able to disperse heat as efficiently if layers of materials like silicone or leather are added on top of a case. As a consequence, your iPhone can get too hot.

Additionally, screen protectors experience this. All screens produce heat, and adding a layer of glass to the top makes the cooling process more difficult and increases the risk of your iPhone overheating.

YOU'D LIKE TO GET RID OF WASTE

Every year, we manufacture millions of phone cases, many of which are constructed from plastic and other non-biodegradable materials. Choosing not to purchase a case reduces waste and safeguards the environment. Case purchases will decrease if enough customers decide not to make them, which is beneficial for the environment.

YOU WANT THE FASTEST WIRELESS CHARGING

Some covers are thicker than typical, and if they're built with the wrong materials, that might interfere with your iPhone's wireless charging. Although you can remove the cover if you wish to charge your phone wirelessly, this isn't a major issue, but you should be aware of it. If you don't utilize a MagSafe-compatible casing, any MagSafe accessories you have can likewise cease functioning.

Naturally, using a high-quality or official case won't interfere with wireless charging or MagSafe accessories.

MAKE THE MOST OUT OF YOUR IPHONE

It only makes sense to want to maximize the use of a gadget when you spend a lot of money on it. Using a cover and screen protector may prevent you from fully appreciating the design of your iPhone, but it may also ensure that it lives as long as possible.

Although going without a cover or screen protector could make your experience better, it definitely isn't worth it if it makes you worry about your iPhone all the time.

ARE IPHONE 13 CASES COMPATIBLE WITH THE IPHONE 14?

Every year when new iPhones are released, Apple also offers a variety of iPhone accessories to match your purchase. However, to avoid spending extra money on new covers if you're upgrading from an iPhone 13, you may want to see whether your current cases work with the iPhone 14.

Here, we'll examine if your old iPhone 13 cases will fit the iPhone 14 models, discuss why Apple continues altering the iPhone's size, and assess whether a cover is really necessary. Let's get going.

CAN YOU USE IPHONE 13 CASES ON THE IPHONE 14?

The case compatibility of the iPhone 13/14, iPhone 13 Pro/14 Pro, and iPhone 13 Pro Max/14 Pro Max/14 Plus will be compared. This year, there is no iPhone 14 small; instead, Apple has introduced the significantly larger iPhone 14 Plus.

Dimensions of the iPhone 13 is 146.7 x 71.5 x 7.65 mm.

Dimensions of the iPhone 14 is 146.7 x 71.5 x 7.8 mm.

The differences between the iPhone 13 and 14 are so negligible that they might almost be considered the same device. The iPhone 14's slightly larger camera module, however, is what renders the two smartphones' case compatibility unusable. Although not perfect, your iPhone 13 cover will work with the iPhone 14. Even though it will be practical, it won't look attractive.

Dimensions of the iPhone 13 Pro are 146.7 x 71.5 x 7.65 mm.

Dimensions of the iPhone 14 Pro are 147.5 x 71.5 x 7.85 mm.

You will without a doubt need a new cover for the iPhone 14 Pro. The placement of the power and volume buttons has also been altered, in addition to the larger camera module. Therefore, the iPhone 14 Pro will not fit your existing iPhone 13 Pro cover.

Dimensions of the iPhone 13 Pro Max are 160.8 x 78.1 x 7.65 mm.

Dimensions of the iPhone 14 Pro Max are 160.7 x 77.6 x 7.85 mm.

Dimensions of the iPhone 14 Plus are 160.8 x 78.1 x 7.8 mm.

The iPhone 14 Pro Max or the iPhone 14 Plus will not fit in your iPhone 13 Pro Max case. The buttons are now in a different place on the iPhone 14 Pro Max, which also sports a larger camera module. Although the buttons on the iPhone 14 Plus are still in the same area, the smaller camera module makes it seem out of place in an iPhone 13 Pro Max case.

DO YOU NEED A CASE FOR THE IPHONE 14?

Like its predecessor, the iPhone 14 series is very durable and includes a Ceramic Shield coating on the front glass; according to Apple, this makes it the world's strongest smartphone glass ever. You can relax knowing that your iPhone can withstand a fall with ease.

Despite this, we still advise purchasing a case and screen protector since the likelihood that your phone's glass may fracture relies more on the angle at which it falls than the surface.

This little change in design is unpleasant if you replace your iPhone every year since it necessitates buying a new cover each time, increasing your carbon footprint.

You may counter that many who retain their iPhones for several years eventually purchase new cases because the old ones get stained or broken.'

WHY NEW IPHONES DON'T FIT OLD IPHONE CASES

Apple continues tweaking the appearance of the iPhone ever-so-slightly every year for one simple reason: to compel you to purchase new MagSafe covers. This is in addition to the necessity for additional space on the smartphone to accommodate physically bigger cameras.

For Apple, this implies that any customer who purchases an iPhone will likely also purchase a new case, ideally from the official website, increasing Apple's income. Most consumers like to put a cover on their phone.

PURCHASE A FRESH IPHONE CASE FOR THE IDEAL FIT.

Purchasing a case designed expressly for your new iPhone is the only way to ensure that it fits inside of it without any gaps or distortions. While you may theoretically use your iPhone 13 cover on the iPhone 14 without any problems, it would be wiser to simply get a new case for the Pro/Plus models.

Keep in mind that you don't have to spend a lot of money on a costly case from Apple; many third-party iPhone 14 covers provide comparable protection and can be purchased for much less on Amazon.

Scan the QR code to get the free Apple Watch Manual

CONCLUSION

If you're an iPhone lover, you might want to make your phone run longer and more effectively.

You may or may not be aware of a variety of maintenance instructions for your iPhone that will help you make the most of it for the duration of its useful life.

To keep your iPhone running longer and more effectively, keep reading for crucial maintenance suggestions.

1. PROTECTIVE CASE

Get a protective case and a screen protector for your iPhone as soon as possible to ensure that you are keeping up with it. Whether you have a brand-new iPhone or not, this should be your top priority.

There are numerous options for protective iPhone cases. When it comes to cases, you may always purchase your preferred brand or style. Finding the best case to match your personality shouldn't be difficult.

2. CHECK THE CLEANLINESS OF YOUR CHARGING PORT

You should make sure that your charging port is clean because charging your phone could be one of the most crucial activities you perform. A clogged lightning port may occasionally be the reason for your phone's inability to charge.

To start with, clean up the junk in your charging port with a toothpick. Don't harm any of the contact sites, therefore proceed with extreme caution. You can try using a can of compressed air to remove any debris if that doesn't work.

3. DELETE APPS YOU AREN'T USING

There are undoubtedly a lot of apps on your phone that you downloaded in the past but haven't used yet. You never check social media, dumb games, or anything else. You should periodically check to see which apps you aren't utilizing and eliminate them when it comes to maintaining your iPhone.

Even more choices can be found in your iPhone Storage settings under the heading "Offload Unused Apps." When you run out of storage, this feature will automatically uninstall useless apps while preserving your documents and data. You can always re-download that program from the app store if you decide at some point that you need it on your phone once more.

4. CONSISTENTLY UPDATE YOUR IOS DEVICE

The best approach to care for your iPhone is to keep your iOS updated. Apple publishes bug fixes and battery life enhancements that can be downloaded through the most recent version of iOS.

It is quite simple to perform and should only take a few minutes. Never neglect to update your iPhone because that is when issues will start to appear.

5. UPDATE YOUR APPS

You should make sure that all of your used apps are updated after uninstalling all of the apps you aren't using, as we've already discussed. By updating programs you don't use, you don't want to eat up internet or time.

There is a function in your settings that will update your apps for you if you frequently forget to do so. As it runs in the background, this will help to guarantee that your apps are always current and ready to use. However, a lot of people turn this off because it can drain the battery.

6. CONTROL BACKGROUND JOBS

On your iPhone, many applications will update in the background while you're doing something else. Your battery life may suffer from this, and it's not always necessary.

When you aren't using the app, the "background app refresh" feature will check for updates. You might wish to do this with some programs, such as your email. For any one of your apps, you can enable or disable the feature.

7. RESTART YOUR IPHONE

You may want to think about resetting your iPhone if you've done everything and it still isn't functioning properly. This may assist in resolving unforeseen problems that you would not have been able to resolve.

8. DELETE ALL SAFARI DATA AND COOKIES

By deleting the cookies and other data utilized by safari, you may improve the performance of your iPhone. You may delete it quickly by heading to your settings, selecting Safari, and then selecting "clear website data and history." Safari will, however, stop recommending websites that you often visit until you save them or return to them later if you do this.

9. DELETE OLD MESSAGES

Some individuals may find this difficult, but you don't need all of the texts you've exchanged with your pals over the course of several months kept on your phone. By enabling the function inside your phone's storage settings, you may effortlessly delete outdated texts.

You may also swipe left on each message to delete it if you simply want to get rid of a few of them. Because they take up the most space, communications with images, videos, or gifs should be your main emphasis.

10. ALWAYS CHARGE YOUR IPHONE

Keeping your iPhone charged so you can use it may seem simple, but this is a great technique to help you conserve battery life over time. The battery will discharge more quickly the more you use your gadget.

It is advised that you always maintain your phone charged, ideally between 40% and 80%. To meet this demand, you should modify how often you recharge your phone.

11. MAKE SURE IT'S SAFE

Keep your iPhone secure, particularly while you're not using it, as the last maintenance advice for keeping it functioning longer and more effectively. Don't leave it someplace where someone may accidentally knock it over or pour anything on it. Keeping your phone secure and out of harm's way is the best method to maintain its functionality.

THE IMPORTANCE OF IPHONE MAINTENANCE

There are a lot of things you should do to maintain your iPhone in order to keep it functioning as long as feasible and as effectively as possible. Maintaining your iPhone is one area where you shouldn't let up.

Do Not Go Yet; One Last Thing To Do

If you enjoyed this book or found it useful, I'd be very grateful if you'd post a short review on Amazon. Your support does make a difference, and I read all the reviews personally so I can get your feedback and make this book even better.

Thanks again for your support!

Made in United States
North Haven, CT
21 February 2023

32966067R00089